Elements
of Mediation

Elements
of Mediation

Sharon C. Leviton, Ph.D.
Adjunct Professor of Law
Texas Wesleyan University School of Law
Diplomate, ABECI

James L. Greenstone, Ed.D., J.D.
Adjunct Professor of Law
Texas Wesleyan University School of Law
Diplomate, ABECI

Brooks/Cole Publishing Company

I(T)P ® An International Thomson Publishing Company

Pacific Grove • Albany • Belmont • Bonn • Boston • Cincinnati • Detroit • Johannesburg • London •
Madrid • Melbourne • Mexico City • New York • Paris • Singapore • Tokyo • Toronto • Washington

 A CLAIREMONT BOOK

Editor: *Eileen Murphy*
Editorial Assistant: *Lisa Blanton*
Interior Design: *Vernon T. Boes*
Cover Design: *Katherine Minerva*
Marketing Team: *Deborah Petit & Jean Thompson*

Production Editor: *Mary Vezilich*
Typesetting: *CompuKing*
Printing and Binding: *Malloy Lithographing, Inc.*

I⊕P The ITP logo is a registered trademark under license.

For more information, contact:

BROOKS/COLE PUBLISHING COMPANY
511 Forest Lodge Road
Pacific Grove, CA 93950
USA

International Thomson Editores
Seneca 53
Col. Polanco
México, D. F., México C.P. 11560

International Thomson Publishing Europe
Berkshire House 168-173
High Holborn
London WC1V 7AA
England

International Thomson Publishing Japan
Hirakawacho Kyowa Building, 3F
2-2-1 Hirakawacho
Chiyoda-ku, Tokyo 102
Japan

Thomas Nelson Australia
102 Dodds Street
South Melbourne, 3205
Victoria, Australia

International Thomson Publishing Asia
221 Henderson Road
#05-10 Henderson Building
Singapore 0315

Nelson Canada
1120 Birchmount Road
Scarborough, Ontario
Canada M1K 5G4

International Thomson Publishing GmbH
Königswinterer Strasse 418
53227 Bonn
Germany

Printed in the United States of America
10 9 8 7 6 5 4 3 2 1

Library of Congress Cataloging-in-Publication Data
Leviton, Sharon.
    Elements of mediation / Sharon C. Leviton, James L. Greenstone.
        p.   cm.
    Includes bibliograpical references (p.   ).
    ISBN 0-534-23982-X
    1. Mediation.   2. Negotiation.   3. Conflict management.
    I. Greenstone, James L.   II. Title.
    BF637.M4L38   1996
    308.6´9—dc20   96-31849
                                                                          CIP

Dedicated to the time when litigation will be the alternative form of dispute resolution, and to those who will make it so.

Sharon C. Leviton *is an Adjunct Professor of Law at Texas Wesleyan University School of Law where she teaches Family Law Mediation, and she is a member of the faculty of the Attorney–Mediator Institute of Houston, Texas. She is a family dispute mediator and crisis specialist in private practice in Dallas, Texas. She was one of the first mediators and trainers for the Dispute Mediation Service of Dallas. She is a Practitioner Member of the Academy of Family Mediators and served as one of the organization's early training supervisors.*

Dr. Leviton is vice chairman of the American Board of Examiners in Crisis Intervention, executive director of the Southwestern Academy of Crisis Interveners, and a fellow of both the American Academy of Crisis Interveners and the Southeastern Academy of Crisis Interveners.

Dr. Leviton is the coauthor of Winning Through Accommodation: The Mediator's Handbook; Elements of Crisis Intervention; Hotline: A Crisis Intervention Directory; The Crisis Intervener's Handbook; *and* Crisis Intervention: Handbook for Interveners. *She has edited the* Crisis Intervener's Newsletter *and the journal* Emotional First Aid. *Audiocassette series include* Crisis Management and Intervener Survival; Stress Reduction: Personal Energy Management; *and* Training the Trainer. *She has contributed chapters to* Innovative Psychotherapies *(edited by Raymond Corsini),* The Wiley Encyclopedia of Psychology, *and* The Crisis Intervention Compendium *(edited by W. Rodney Fowler) and* Mediation Quarterly. *She has published articles, papers, and editorials in family and marital dispute mediation, psychotherapy, crisis intervention, and stress management.*

James L. Greenstone *is the Fort Worth, Texas, Police Department Psychologist. He is an Adjunct Professor of Law at Texas Wesleyan University School of Law where he teaches Family Law Mediation, and is a member of the faculty of the Attorney–Mediator Institute of Houston, Texas. Since 1966 he has been a marriage and family psychotherapist and a family dispute mediator in private practice in Dallas, Texas. Dr. Greenstone is national vice president of the American Academy of Crisis Interveners and the chairman of the American Board of Examiners in Crisis Intervention. He served as president of the Southwestern Academy of Crisis Interveners and editor-in-chief of* The Journal of Crisis Intervention. *Dr. Greenstone is a certified Master Peace Officer in the state of Texas and is the lead hostage and crisis negotiations instructor at the North Central Texas Regional Police Academy.*

Dr. Greenstone serves as editor-in-chief of The Journal of Police Negotiations, Crisis Management and Suicidology, *and is the former editor-in-chief of* The Journal of Crisis Negotiations. *He has been the senior editor of the* Crisis Intervener's Newsletter *and editor-in-chief of the journal* Emotional First Aid. *He is the author or coauthor of* Winning Through Accommodation: The Mediator's Handbook; Elements of Crisis Intervention; The Crisis Intervener's Handbook; Hotline: A Crisis Intervention Directory; Crisis Intervention: Handbook for Interveners; *and* The Crisis Intervention Compendium. *Audiocassette series include* Crisis Management and Intervener Survival; Stress Reduction: Personal Energy Management; *and* Training the Trainer. *Dr. Greenstone has also contributed chapters to* Innovative Psychotherapies *(edited by Raymond Corsini),* The Wiley Encyclopedia of Psychology, *and* Police Psychology into the Twenty-First Century. *Dr. Greenstone's professional memberships include the Academy of Family Mediators, the Society of Professionals in Dispute Resolution, and the Alternative Dispute Resolution Sections of the American and the Texas Bar Associations.*

PREFACE

The Elements of Mediation is intended for the following groups of people.

Mediators
Arbitrators
Attorneys
Counselors
Marriage and family therapists
Psychotherapists
Teachers
Students
Landlords
Tenants
Employers
Employees
Parents
Government agencies
Community agencies
Advocates
Management
Police officers and supervisors
Professional associations
Unions
Real estate agents
Professional and lay people who wish to promote cooperative
    problem solving

We have reduced the practice of Mediation to its basic elements so they can be applied as broadly as possible, and we present them in a format that is useful for the experienced professional and for the novice. We have spent much time in our careers eliminating confusion about mediation procedures and about the relationship between Mediation and other dispute resolution options. Although mediators often hold licensures in other disciplines, such as law or counseling, their credentials as trained mediators is what matters foremost to the disputants. This book reinforces the theoretical framework that postulates Mediation as a viable discipline in itself.

Because this is a practical guide, most theory has been purposely omitted. We suggest that *Elements of Mediation* be used as a supplement in classes in Mediation, law, dispute resolution, negotiations, counseling, social work, marriage and family, sociology, business, government, police science, communications, education, management, and in related and applied courses at all levels. It is also appropriate for training courses in Mediation and dispute resolution. The experienced mediator can use the book independently, in the classroom, and in the practice.

How to Use This Book

This book is a practical approach to Mediation. It is designed to provide the mediator with concrete and effective techniques to assist parties in dispute resolution. In addition to listing the areas covered in the chapters, the table of contents is a step-by-step guide to the mediation process and should be used by the mediator to guide the Mediation in an orderly fashion. For the experienced mediator, the table of contents is a helpful reminder of the areas to be covered during the mediation process. Novices may need to read the entire book carefully before they can use the table of contents effectively, and they should understand that the full value of this book depends on their gaining theoretical depth and practical training.

Brief exchanges between "mediator" and "party" accompany much of the information to illustrate the element being discussed. Material is frequently presented in list form to produce a user-friendly, easy reference.

In addition to using the prescriptive table of contents, the mediators can also look up material according to the activities they want to perform. This listing is printed on the inside of the back cover of the book.

The Elements of Mediation

We believe that if parties to a dispute can confront the problem together in a good-faith effort, they will discover a solution that is both equitable and consensual. Further, when they have equal responsibility for developing the outcome, the likelihood of performance increases dramatically.

The mediation process provides a mechanism for parties to reach a settlement through nonjudicial, nonadversarial proceedings. Its great strength is in its inherent mandate to return the opportunity and the responsibility for the conflict resolution to the people directly involved. The process has the capacity to

- ease the load on the court docket.
- reduce the expense of settlement.
- lessen the degree of stress and tension experienced by the disputants.
- provide the possibility of an outcome that meets the needs of both parties.
- provide an instructive experience to aid parties involved to proceed with their lives more efficiently and productively.
- provide a setting for disputants to be the author of the settlement rather than the victim of imposed judgment.

Currently Mediation is considered an alternative process to the primary form of conflict resolution—litigation. Our dream is that one day in the not-too-distant future, Mediation will be recognized as the norm with litigation becoming one of the several alternatives. This will only happen if the public views Mediation and the mediator as being effective. The keys to mediator effectiveness are a clear understanding of the elements of Mediation and the ability to apply them practically.

We would like to acknowledge the helpful comments and suggestions of the book's reviewers: Susanne C. Adams, The Mediation Group, Inc.; Richard Evarts, Settlement Consultants International, Incl.; W. Rodney Fowler, University of Tennessee at Chattanooga; James R. Mahalik, Boston College; Mary Finn Maples, University of Nevada; Peter E. Maynard, University of Rhode Island; Patricia McCarthy, University of Minnesota; Michael J. McMains, San Antonio Police Department; Jane R. Rosen-Grandon, Rosen Grandon Associates, Inc. Finally, our thanks to the people at Brooks/Cole: the editorial assistant, Lisa Blanton; the production editor, Mary Vezilich; the editor, Eileen Murphy; and Claire Verduin, without whom this book would not have been possible.

Sharon C. Leviton
James L. Greenstone

CONTENTS

h. *Balance personal need for control with parties' need to discuss issues with each other*
i. *Monitor personal feelings regarding the conflict*
j. *Maintain fairness*
k. *Open communications channels*
l. *Translate and transmit information*
m. *Help distinguish between wants and needs*
n. *Create options*
o. *Serve as an agent of reality*
p. *Help the parties to save face*
q. *Recognize and respond to mounting tensions*
r. *Recognize and respond to withdrawal of either party*
s. *Help to break impasses*
t. *Be sensitive to ethical considerations*
u. *Stay within the boundaries of the process*
v. *Set the time and place for Mediation*
w. *Insure privacy for all during the sessions*

CHAPTER THREE
Preparing the Physical Setting 17

1. Choose a neutral site
2. Plan and arrange the office setting appropriately
3. Plan furniture and seating arrangements carefully
4. Use information gained from the referral source, the intake, and the "field trip" in planning the setting

CHAPTER FOUR
Beginning the Mediation 22

1. Bring the clients into the mediation room
2. Seat the parties
3. Deal with the feelings of the clients, as needed
   a. *Be aware of brewing or overt hostility*
   b. *Be aware of unusual concerns or of withdrawal*
   c. *Be sensitive to any behavior that seems inappropriate or unusual*

4. Make the environment safe to negotiate
5. Make an opening statement
   a. *Introduce yourself*
   b. *Establish yourself as credible and open*
   c. *Obtain the names and identify the roles of all persons in the room*
   d. *Disclaim bias or partiality*
   e. *Explain mediator's role and authority*
   f. *Emphasize the informal nature of the process*
   g. *Review the mediation process*
   h. *Outline the procedures to be followed*
      1) Designate the order of initial presentation
      2) Set the ground rules
   i. *Explain the caucus and when it might be used*
   j. *Discuss confidentiality and privilege*
   k. *Clarify parties' ability and authority to enter into an agreement*
   l. *Deal with the logistics*
   m. *Clarify parties' time commitments*
   n. *Clarify parties' willingness to mediate in good faith*
   o. *Answer questions and respond to concerns about the process*
   p. *Get a commitment to begin*

C H A P T E R   F I V E

Ventilation, Information Gathering, Problem Solving, and Bargaining   31

1. Allow for ventilation to occur
2. Gather information about the dispute
3. Clarify issues
4. Establish negotiating distances
   a. *Identify what the parties want*
   b. *Identify what the parties need*
   c. *Establish the distance between the wants and the needs*
   d. *Remain interest based rather than position based*
5. Generate options and alternatives
   a. *Focus on needs*
   b. *Discover ways of meeting the needs of all parties*

CHAPTER NINE

Ethical Considerations in Mediation 68

1. Reveal conflicts of interest
2. Reveal previous relationships
3. Charge a fair fee
4. Communicate confidentiality and privilege
5. Communicate the rules of Mediation
6. Advise that mediator is not a witness for either side
7. Deal with contacts between all parties involved
8. No subpoenas will be served immediately before, during, or immediately after a mediation session
9. Allow for problem ownership by the disputants
10. Avoid giving personal opinions

CHAPTER TEN

Dealing with Hostility 70

1. Review guidelines for understanding and for handling hostility or hostile gestures in Mediation

CHAPTER ELEVEN

Stressors, Stress Management, and Crisis Intervention 72

1. Be aware of the potential for heightened client stress in Mediation
2. Take a proactive approach to minimize client stress
   a. *Recognize the stressors related to Mediation*
   b. *Screen the parties and the case*
   c. *Arrange the office carefully*
   d. *Take a field trip through the waiting room*
   e. *Plan the opening statement*
   f. *Caucus as appropriate*
   g. *Empower the parties*
   h. *Remain aware and sensitive*

3. Recognize maladaptive behavior
4. Know how to intervene in a crisis
5. Move from Mediation to crisis intervention
6. Abort the mediation session, if indicated
7. Move back to Mediation, if possible

CHAPTER ONE

Approach to Mediation

Mediation is the intervention of a neutral third party, who intervening at the request of the parties, assists the parties at dispute in finding their own way out of the dispute through equity through consensus (Evarts, 1980). A less formal description of Mediation is that it is a process in which a neutral third party, called a mediator, helps disputing parties arrive at a mutually acceptable solution to their conflict. Mediation is a step-by-step process in which agreement and disagreement are carefully explored, relevant information is collected and shared, options and proposals are discussed, interests of each party are presented and clarified, and negotiations between the parties are conducted to resolve the conflict. The decision-making power and responsibility remain with the parties. The mediator acts as a neutral facilitator (Zaidel, 1990).

The following are key elements in the above definitions of Mediation:

- The intervener is not invested in the outcome of the Mediation.
- The intervener serves as a neutral party.
- A dispute exists between at least two parties who wish to resolve the conflict through a consensual agreement.
- The intervener is prepared to facilitate the mediation process to assist the parties in achieving a consensus-based resolution to their conflict.
- The intervener has no coercive powers. The goal of Mediation is the construction of an equitable settlement based in consensus.

Other Interventions

Other conflict intervention processes are sometimes confused with or mislabeled as Mediation. Arbitration, conciliation, litigation, psychotherapy, and crisis intervention must be differentiated from Mediation.

• *Arbitration* may be defined as follows: "The reference of a dispute to an impartial (third) person chosen by the parties to the dispute who agree in advance to abide by the arbitrator's award issued after a hearing at which both parties have an opportunity to be heard" (Black, 1983). "Arbitration is a process which is utilized when the parties voluntarily agree, in writing, to resolve disputes which may arise out of a contract. The arbitrator, a neutral person, presides at the hearing, hears all the facts and evidence of the parties, and thereafter renders an award which is final and binding. The award may be entered as a judgement (sic) of the court" (H. O. Wolff, personal communication, September 17, 1993). Unlike Mediation, arbitration is judgment based.

• *Conciliation* is an informal process in which a third party tries to facilitate an agreement through separate consultation with disputants. Conciliation is more typically used to describe a process that occurs before the parties are ready to commit themselves to formal Mediation. Conciliators attempt to reduce tension, clarify issues, and get the parties talking.

• *Litigation* is a process in which a judge is authorized to interpret and apply the law on matters brought before the court. In litigation, the parties are represented by opposing counsel, each of whom seeks to establish a clear "victory" for his or her client. Litigation is judgment based.

• *Psychotherapy* is a process of working with patients to assist them to modify, change, or reduce factors that interfere with effective living. Counseling and psychotherapy begin from a premise of what constitutes health and then probes deeply into the motivations of behavior.

• *Crisis intervention* may be defined as the timely intrusion into the life of a person who, because of unusual stress, cannot handle life as usual (Greenstone, & Leviton, 1993). The crucial issue is heightened stress and its toll on effective functioning. At such times, Mediation, which requires rational thinking and involvement, is not possible. Crisis-level stress may not only leave the party unable to deal effectively with the dispute, but also produce maladaptive behavior. Crisis recognition and crisis intervention in Mediation are vital. Stress and crisis intervention skills are discussed in chapter 11 of this book.

Clearly, Mediation is not a judgment-based intervention, as are arbitration and litigation; nor is it counseling or crisis intervention.

Comparing Arbitration and Mediation

Arbitration	Mediation
Formal — rules of evidence	Informal — its strength
Witnesses	Generally, no witnesses
Introduction of evidence	Virtually no evidence
Privacy	Confidential (key to the process)
Joinder of parties only by agreement, court order, or contractual	Joinder of parties not unusual
More time	Less time
More expensive	Less expensive
Some venting	Venting
Role of neutral: Decision maker	Role of neutral: Facilitator
Authority rests with decision maker	Authority rests with parties
Filing of briefs	No briefs filed
Meetings are joint	Meetings: Joint and caucus expected
Court reporter allowed	No court reporter allowed
Who is wrong or right?	Let's find a solution everyone can live with.
Award is binding. Award generally due 30 days after hearings closed.	No award. Agreement made by parties, often immediately.

Source: American Arbitration Association, 1440 Two Galleria Tower, 13455 Noel Road, Dallas, Texas 75240-6620

Divorce Mediation is not marriage counseling. Mediation cannot occur if a disputant is in crisis. Mediation is not the practice of law. Attorneys who are also mediators can practice Mediation. Nonattorneys may also be mediators. Although people who have preexisting primary professions may also be mediators, those without preexisting professions may have Mediation as their primary profession.

Referrals from Mediation to Other Processes

Some circumstances require the cessation of Mediation in favor of another intervention process. Examples of such instances follow.

Some Differences Between Arbitration and Litigation
(Arbitration under the Rules of the American Arbitration Association)

	Arbitration	Litigation
COMMENCEMENT	By Demand for Arbitration for future disputes. By Submission Agreement for existing disputes.	File suit—no permission needed.
APPOINTMENT OF DECISION MAKER	Select arbitrator according to Rules. Strike unwanted arbitrators. Number those acceptable in order of preference.	No opportunity to to choose judge.
ADMISSABILITY OF EVIDENCE	No formal rules. Appropriate weight given.	Formal rules.
ON-SITE INSPECTION	Possible.	Generally cannot be done.
SPEED	Expedited; can be heard quickly. Regular cases about 4 to 6 months to complete. Average case should not take more than 2 days of hearing.	Can take years to be heard. Appeals take longer.
EXPERTISE	Arbitrators with specialized knowledge. Over 50,000 arbitrators available from U.S. and throughout the world.	Vast expertise generally not available.
COST	Fee varies—depending on amount in dispute. Generally less costly than litigation.	Generally more expensive.
APPEALS	Governed by statute—final and binding. Cannot appeal on merits. Can appeal if arbitrator has exceeded power, fraud, misconduct, miscalculations, and other basis per statute.	Can appeal.
DISCOVERY	Discouraged but is possible.	Used extensively.
SCHEDULING	At convenience of parties and arbitrators. Firm date set.	May not be heard when scheduled. Attorneys need to appear for docket call.
PRIVACY	Only those connected with dispute generally attend hearing.	Open to the public and press.
PLACE OF HEARING	Most anywhere, sometimes at site of idspute. Not required to be held at AAA Regional Office.	Courtroom.

(continued)

(continued)

	Arbitration	Litigation
USE OF ATTORNEYS	In 9 of 10 substantial cases, an attorney represents the client at hearing.	Attorneys used.
FILING OF BRIEFS AND USE OF COURT REPORTER	May be done.	May be done.
USE OF EXPERT WITNESSES	Yes, but not always necessary, as arbitrators have expertise.	Yes.
DECISION	Rendered 30 days from close of hearing. If expedited, 14 days from close of hearing. (AAA Rules)	May be an extended period of time.

Source: American Arbitration Association, 1440 Two Galleria Tower, 13455 Noel Road, Dallas, Texas 75240-6620

Mediation + Arbitration

Jack Brown and Joe Green reached consensus on three of the issues that they had brought to Mediation. They could not mutually resolve the final problem and decided to ask an arbitrator to settle it. So, a combination of Mediation and arbitration was used to resolve their dispute.

Mediation + Counseling

During a divorce mediation session to consider custody and visitation of the couple's two children, Mr.. Smith had blocked every attempt by his wife to negotiate with him. His unresolved bitterness and deep resentment toward Mrs. Smith for initiating the divorce interfered with his willingness to work with her. After attempts to move Mr. Smith out of the venting stage, the mediator stopped the session and suggested that Mr. Smith deal with the anger and grief outside of the Mediation in therapy or supportive counseling. Until he could make the separation between feelings and behavior, Mediation was not a viable option for them. A person can be in psychotherapy while also being a party to Mediation. The key element in determining suitability is the extent of the emotional or psychological impairment.

Litigation

Sally Young admitted to what the mediator had begun to sense. She wanted her day in court. She was convinced that the "judge would give her a better deal than she could get in Mediation." She wanted an advocate to plead her case. Mediation ended.

Determine Whether the Parties Are Appropriate for Mediation

Mediators will be called by well-meaning therapists, attorneys, potential clients, and even other dispute resolution professionals to mediate cases that are not appropriate for the mediation process. The following are examples of cases that are not appropriate for the mediation process.

Attorney: I am looking for a mediator who is a family therapist. I understand that you fit that description, and I want to refer a couple to you.

Dr. Leviton: Are you referring them for therapy or for Mediation?

Attorney: Well, I want the husband of my client to be seen by the court as unfit to have joint custody of their children. When they come to Mediation and he blows up and becomes abusive, you can tell him that he needs counseling. He will either get help at best, or I can report to the court that the mediator said the wife came to Mediation in good faith.

Dr. Leviton: The case you are describing is not appropriate for Mediation. The agenda is antithetical to the mediation process.

A couple is being briefed on the mediation process. Mary Jones is tentative in her questioning. She defers to John Jones, her husband, before responding to the mediator. Mr. Jones is ebullient in his praise of Mediation and how the agreement that they will reach will improve their lives.

Mediator: I will ask that you provide a full accounting of all your assets for the two of you to examine jointly.

Mrs. Jones: Does that mean the farm property that I know exists but really have no details about?

Mr. Jones: No. Don't worry about that. It has nothing to do with you.

Mrs. Jones: What about that boathouse?

Mr. Jones: I told you those things have nothing to do with you. They are not part of Mediation.

Mediator: The mediation process requires disclosure of all relevant information as would reasonably occur in the judicial discovery process. You both need equal access to information in order to negotiate intelligently.

Mrs. Jones: We have never looked at our financial matters together. He has always been secretive about everything.

Mr. Jones: Maybe we had better rethink this mediation thing. I have another appointment in a few minutes anyway. I'll call you if we decide to come back.

Mediation is concerned with consensual resolution rather than imposed settlement. It requires that the parties to the dispute take responsibility for their own actions, negotiations, contractual arrangements, and individual performance during and after the fact. They must make a commitment to participate in good faith with the intention to settle, if at all possible. The process requires that the mental, emotional, physical, and intellectual capacities of the parties are equal to the task of full participation.

The following are signals to the mediator of trouble in this regard:

- Party expects the mediator to resolve the conflict for them.
- Party expects Mediation to be psychotherapeutic. In a divorce mediation case, party expects divorce counseling or marriage counseling.
- Party wants retribution. The concept of interest-based negotiations is foreign to his or her agenda.
- Party wants his or her position vindicated by a third party, i.e., the mediator. Party will try to seduce the mediator into siding with him or her.
- Party believes that through the informal nature of the process he or she can circumvent disclosure of something he or she doesn't want revealed.
- Party appears at the session intoxicated.
- Party exhibits maladaptive behavior.

- Party displays a tenuous grip on reality.
- Party plans to use Mediation as a delaying tactic to avoid settlement.
- Party displays violent or hostile behavior.

During the intake, mediators should be alert to what they are hearing, seeing, feeling, and sensing in the words and behavior of the parties. Careful attention at the outset will prevent problems later on.

CHAPTER TWO

Prepare to Mediate

This chapter describes key elements in preparing for a mediation session. After a brief view of when and how cases come to Mediation, emphasis will shift to the importance of preparing for the cases and the parties.

When Do Parties Choose Mediation?

The following circumstances are particularly appropriate for Mediation:

- The parties' different perceptions of the problem prevent them from moving beyond the argument.
- The parties have tunneled their vision to the extent that the only possible solution they see currently is the position they have adopted.
- The parties are pressured by deadlines and time constraints and are anxious for an early settlement. Mediation can often be arranged within a short time of the initial contact and concluded in a single session or several sessions arranged close together.
- The parties recognize that they must rely on mutual cooperation to meet their goals or satisfy their interests.
- The parties don't want to go to court or have their dispute settled in a public forum.
- The parties no longer believe that they are able to manage the conflict on their own. They recognize the need for an impartial third-party intervention.
- The parties want to resolve the conflict themselves, but they lack the negotiation skills to do so. They want the assistance that a mediator can provide.
- The parties recognize the benefit in maintaining a positive

future relationship. Examples include family conflicts, neigh-borhood disputes, and partnership disputes.

• The interests of all parties are mutually interdependent. Ex-amples include family conflicts, public-interest disputes, en-vironmental disputes, neighborhood disputes, and partner-ship disputes.

• The parties seek solutions that will maximize gains and mini-mize losses for both sides.

The term a *win-win outcome* is often attributed to the benefit of using the mediation process. A win-win outcome occurs when both parties to the dispute feel that their interests have been sufficiently satisfied.

How Do Cases Come to Mediation?

Cases are initiated in the following ways:

• By the parties' self-referral.

• By referral from the attorney representing the party. In this situation, the attorney might be involved in the Mediation as an advisor on legal matters, as a negotiator for the client, as a silent observer, or in some mix of the above. The media-tor cannot give legal advice and therefore would ensure that the parties seek counsel on legal issues during the negotia-tions and again in drafting a final agreement to be presented to the court.

• By referral from other resources who have a professional re-lationship with the party. Examples include therapist, phy-sician, accountant, and clergy.

• By referral from the court. In some states the court will order Mediation in particular cases.

• As a result of a contract that includes a mediation clause should a dispute arise in connection with the performance of the terms of the contract.

Who Is Eligible to Be a Mediator?

In principle, anyone can function as a mediator provided he or she is impartial and trusted by both parties. Informal Mediation for all kinds of disputes has been in existence since civilization began (Evarts, Greenstone, Kirkpatrick, & Leviton, 1983). The use of Mediation has been constantly expanding into international, interpersonal, and interorganizational affairs. This increased interest in Mediation in recent years has created an impetus for formalized training of mediators. Concerns about certification, credentialing, and standards of practice have transformed a previously informal function into a profession (*Texas Civil Practice and Remedies Code*, 1988).

A mediator must have both technical skill and an innate talent. The skills are learned in the 40-plus hours of classroom instruction, in-service training, and practical experience. The talent is reflected in the sensitivity with which the mediator listens, hears, responds, empathizes, creates, draws parties into the process, and deftly maneuvers through the delicate and difficult moments with an intuitive sense of timing and appropriateness.

What Does the Mediator Do?

The Mediator is primarily a facilitator who provides the parties with a process through which there can be a joint examination of the issues at dispute, an identification of common objectives, and insights into opposing perspectives. The mediator makes no judgments as to the merit of positions and renders no decision as to which party shall prevail.

In performing the job, the mediator takes on many roles. These include:

Process Facilitator   allows the discussion to occur by providing a neutral environment, arranging and chairing sessions, establishing and maintaining procedural guidelines for the sessions, and creating a climate conducive to problem solving.

Discussion Facilitator   assists the parties in identifying and clarifying the issues that set the agenda, keeps the disputants on track, and keeps a discussion moving by intervening when necessary to reestablish communication between parties.

Clarifier   when necessary, puts each party's terms into language that the other party can understand. The alert mediator recognizes that often people talk much, hear little, and understand even less of what is being said. Nuances are missed, terms are misunderstood or unfamiliar, meanings are distorted, and/or feelings are misinterpreted. The parties are talking but clearly not understanding each other. They not only are unaware of certain critical facts, but often have different perceptions of the same set of facts. People create their own reality based on their particular background, experience, and observations. They bring their reality into the Mediation and then measure the issues and the possibilities of settlement against that reality. The mediator must continually probe for mutual understanding of the information being presented. The mediator first clarifies his or her own understanding of the content and meaning of what the parties are saying and then, as necessary, acts as an interpreter or translator for the disputants.

Mediator: Mr. Jones, I am not familiar with the terms you're using. Could you explain them to me? I want to make sure that we are understanding each other.

Mediator: Miss Smith, when you talk about "some time in the future," could you be more specific? It would help us to know what time frame you have in mind.

Mediator: Mr. Brown, you said you would pick up the children mid-morning.

Mr. Brown: That's right. Mid-morning. Around 11:00 A.M.

Mrs. Brown: I'm up at 4:30 A.M. to get ready for work. Mid-morning to me is 8:00 A.M. I've already been up 4 hours.

Mediator: Would each of you tell me your understanding of the contract.

Ms. Green: It states that all work on the apartment must get written approval from the rental agent before work commences.

Rental Agent: It clearly states that any changes to the apartment must have written approval from the rental agent before work commences. What could be clearer than that!

Mediator: You both seem to understand the fact that changes to the apartment must have written approval from a rental agent before work can begin. Let's look at where the confusion is coming from.

Ms. Green: I just painted the place a different color. I didn't change anything like moving any walls or tearing through anything.

Rental Agent: Painting a different color is changing . . . .

Idea Generator encourages the consideration of new ideas or alternatives when the parties are stuck or have given up. This can be the fun part of a mediator's role because of the opportunity for creativity. The mediator can draw the parties into a discussion by suggesting hypothetical possibilities, such as

- What would happen if _____?
- Have you thought about _____?
- Have you investigated the possibility of _____?
- How could you develop _____?

IT IS NOT THE MEDIATOR'S ROLE TO CREATE A SOLUTION AND PROMOTE IT. It is the mediator's role, however, to keep the parties engaged in problem solving. Sometimes that means probing ideas and asking questions in an effort to stimulate creativity. The infusion of new ideas can become a catalyst for the generation and consideration of options and possibilities. The mediator would encourage the parties to consider proposals without a commitment to adoption. As long as parties are exchanging ideas, they allow the possibility of discovering a viable solution that appeals to the interests of all involved.

Face Saver suggests a way that helps one or both parties create an agreement that allows them to save face. Often the mediator will hear a party make one of the following statements:

- I really would like to end this battle, but my friends will say I gave in to him.
- My kids will think that I didn't fight hard enough to get custody of them if I agree to that.
- I don't need those things, but I don't want my boss to think I'm a wimp.

Mediator: Mr. Jones, you have rejected the several offers by

> Mr. Kinney. Could you tell me what you really need to settle this?
>
> **Mr. Jones:**   He embarrassed me big time. When all is said and done what I really need to have is an apology from him. I need my family and friends to see a written apology. That would end this whole thing.

Agent of Reality   helps the parties to recognize what is and what is not a practical solution to their differences. The mediator asks each side to think through and justify its facts, demands, and interests; has each party assess the costs and benefits of either resolving or continuing the conflict; asks each party to consider and deal with the other's arguments and interests; questions the benefit of maintaining rigid positions; and explores alternatives. By asking questions that probe to the key issues of the dispute, the mediator helps the parties come to a resolution.

Further discussion of these points will be found in chapter 6 on breaking deadlocks.

Messenger   shuttles between the parties when the circumstances or the level of hostility preclude direct communication.

Distinguisher of Needs from Wants   helps the parties distinguish their true underlying needs from what they merely want. Wants sometimes become disguised as needs. Needs are irreducible elements that must be addressed and satisfied for the dispute to be settled. When people feel hurt, angry, disappointed, or wronged, they often operate initially out of reaction to their misery. At those times, emotion supersedes calm judgment. The mediator must help sort through the substantive issues and the emotional issues to discover what true needs must be satisfied if the dispute is to be settled.

> **Mediator:**   Mrs. Gardener, you have listed keeping the house as a primary need. How will keeping the house be of benefit to you?
>
> **Mrs. Gardener:**   I must have that house!
>
> **Mediator:**   Is it that you have to have that particular house or a house?
>
> **Mrs. Gardener:**   I need that house.

Mediator: Mmmm. You said earlier that you wish you could get out from under so many burdens. You repeatedly mentioned debts and responsibility for the care and maintenance of the house and its several gardens. Mrs. Gardener, do you see where selling the house might ease any of those burdens?

Mrs. Gardener: He thinks we should sell the house and use the money to pay off some of our debts and also create a college fund for Jane.

Mediator: How do you feel about that idea?

Mrs. Gardener: I'm so angry at him about this divorce. Anything that's his idea I don't want.

Mediator: I hear how angry you're feeling. Even though it's Mr. Gardener's idea, how might selling the house be of some benefit to you?

Mrs. Gardener: Well, I need a nice place to live, but I also need a lot of peace of mind. I feel like I'm being pulled in too many directions. Between you and me that beautiful house has become a liability to me. Maybe I can find a smaller place that won't require so much upkeep. John offered to work with me on finding something I'd like.

Trainer   teaches the parties how to negotiate more effectively with each other. This knowledge is of particular importance when the parties will have an ongoing relationship. Examples include family disputes, partnership disputes, neighborhood disputes, and landlord/tenant disputes.

Additional functions of the mediator are described or listed throughout the book. All of the roles are designed to help the parties do the following:

- Communicate with each other.
- Identify and separate substantive issues from emotional issues.
- Understand and appreciate each other's concerns, issues, and interests.
- Reassess their own positions on an issue.
- Recognize when they are in agreement on an issue.

- Recognize superordinate goals—goals that they share and that require mutual cooperation to achieve.

The mediator must be prepared to do the following:

- Probe and ask questions that will provide information.
- Observe what the disputants say and what they do. Often body language and nonverbal behavior become more important sources of information than do verbalizations.
- Recognize and respond effectively to mounting tensions.
- Maintain control of the process and keep the disputants on target. Maintenance of control is discussed further in chapter 4.
- Establish rapport, confidence, and trust in a relatively short time. Some factors that build rapport are clarity in seeking and giving information; a personal presence that reflects stability and calm; appropriate interest in both the parties and the conflict; and an ability to reach out verbally and behaviorally, such as a handshake, eye contact, and attentiveness, in order to bring the parties into the process quickly.

Mediators must remain objective and impartial. They must be reassuring and calm, maintaining a certain steadiness and warmth toward both parties. They must be supportive. The disputants will normally open up if they feel that the mediator is interested in their problem and can be trusted to help with the process. Mediators must be patient and allow the parties time to work through the problem and find the solution. Although solutions may not always be available on the spot, mediators must hold out the realistic hope that solutions are possible. They actively try to help the parties mobilize their own resources to affect a resolution. If the parties believe that the mediator is capable of helping them reach a solution to their dispute, they will be more likely to trust the mediator and the process (Greenstone, Leviton, & Fowler, 1991).

The discussion of preparing for a Mediation will continue in chapter 3.

CHAPTER THREE

Preparing the Physical Setting

This chapter continues to build on the importance of preparation for a Mediation. The discussion in chapter 2 described the many roles that mediators must be prepared to perform in meeting with the parties. The focus shifts now from the mediator to the physical setting of the meeting room.

Realistically, a Mediation can be conducted in an open field that is devoid of enclosed rooms, furniture, flip charts, and other conveniences and accoutrements. The key element in affecting a successful Mediation is the firm grounding that the mediator brings to the process. Understanding what to do, when to do it, why it is being done, and what effect it will have has greater importance than where the event is happening. However, the physical setting can have an impact on facilitating communication, gaining and maintaining control of the argument, reducing or increasing pressure on the parties, and insuring the safety of all those involved.

Some of the following material is presented in list form, because the ideas stand alone without need of elaboration or because the suggestions are Mediator Guidelines for Action.

Choosing a Site

Every attempt should be made to select a neutral site for the Mediation. Conducting the sessions at the office or home of one of the parties or at the office of one of the party's attorney should be avoided if at all possible. The perception of a "home court advantage" for one of the parties could easily contaminate the process.

Arranging the Office

A desirable mediation setting would include the following:

- Enclosed rooms with doors that provide complete privacy
- Sufficient insulation to limit sounds or conversations from another room
- Adequate lighting that is free of glare
- Adequate heating and cooling
- An absence of distractions such as ringing telephones or pagers
- Neutral or nonprovocative wall colors
- Adequate ventilation
- Adequate work space
- Minimal clutter
- Window(s)
- Wall treatments, window treatments, or furniture treatments that do not intrude
- Easily accessible restroom facilities
- Adequate accommodations for the elderly or handicapped
- Adequate security within or surrounding the office building
- A separate room(s) that can be used as a caucus room (caucus is explained and discussed in several later chapters)
- Adequate-size room for the number of disputants

Arranging the Furniture

Table shapes have specific effects on a Mediation. If given a choice of using a round, square, rectangular, multisided, or high or low table, mediators should review their information about the disputants and the case before making a selection. They should factor these considerations into their decision:

- Square tables have equal sides, which suggests an equality between the parties. There is no "head" position. No party, including the mediator, can command a superior seating position. The square table can be valuable with two parties and/or with two mediators. It can enhance the parties' attempts at consensus.
- Round tables are devoid of edges and dividing lines. All seats are the same, vision is unobstructed, and the perception is that of bringing people together and decreasing differences.
- Rectangular tables have head and foot positions and hard, square lines. The mediator who uses rectangular tables should take one of the ends and discourage anyone other than a co-mediator

from sitting at the other end. This can be accomplished by removing the chair from that position and directing the parties to their seats.
• Tables vary in height from about 18 inches to about 30 inches. Smaller tables such as coffee-height tables put fewer barriers between people but make it more difficult to write and lessen safety. Desktop-height tables provide a barrier, adding to the perception of safety and make note taking easier.
• The decision to have no table between the parties is used when there is minimum need to write and maximum need to interact, and where the potential for violence is assessed to be minimal.

Chairs and sofas must also be considered from the perspective of their effectiveness. Although a sofa creates a sense of informality, it poses some logistical problems. The parties are discouraged from facing each other unless they twist themselves around. Writing is difficult, and there are no barriers between the disputants. Facing chairs toward each other encourages parties to talk together. Facing them toward the mediator discourages direct interaction between the parties.

In planning a strategy for the session, mediators should consider the effect that the furniture arrangement and the seating plan might have on the outcome. Where a high degree of hostility exists, the setting should enhance the perception of structure and rules; for example, use a rectangular table. When an obvious power disparity exists between the parties because of differences in experience, knowledge, position, or income, the setting should enhance the perception of equal power; for example, use a round table or if using a square or rectangular table have the parties seated directly across from each other (Evarts et al., 1983).

Looking for Clues

The suggestion was made above that mediators use available information about the parties and the dispute as the basis for arranging the setting. That information can come from several sources, including the referral source and intake. An additional source of information may come from what is sometimes referred to as a "field trip" through the waiting room. Guidelines for this field trip are given on the following pages.

Mediator Guidelines for the Field Trip

Take a field trip through the reception area once the parties have arrived but before they are brought into the mediation room. This very brief "scouting out" can provide invaluable clues for assessment purposes.

Some Questions to Ask Yourself
Are the parties sitting together?
Have they chosen to wait in separate rooms?
Is there interaction between the parties?
What is the quality of the interaction?
Does either party appear unusually anxious?
Does either party exhibit maladaptive behavior?
Is there discernible hostility?
Do you smell alcohol?
Is there a disparity in the way they are dressed that might affect the equity of power?
Do you sense anything unusual that you want to remember for later use?
Do you hear, see, smell, or sense something that might necessitate a premediation caucus?

Use the information you have gathered from the intake and the field trip to

- arrange the mediation room.
- arrange the seating pattern.
- determine the need to address special concerns in the introduction. You might have to address feelings before you can move to substantive issues.
- determine whether or not you will proceed with the Mediation.

You can accomplish this walk-through quickly. It is not a time for socializing or visiting with the parties. To the contrary, avoid giving any impression of bias toward either party.

Concerns about client stress and its impact on safety considerations is discussed as separate topics in chapters 11 and 12.

The following is a recap of mediator guidelines for the intake of clients:

- Assess mental capacity for Mediation.
- Assess emotional stability.
- Assess level of hostility.
- Confront the issue of violence, as needed.
- Determine the suitability of issues at dispute for the mediation process.
- Explain the elements of the mediation process to the clients.
- Review with clients information about fees, methods of payment, scheduling of sessions, and basic requirements.
- Answer questions and address clients' concerns about the process.
- Sign an agreement to mediate.
- Complete other appropriate forms, as necessary.
- Set the tone for Mediation to occur.
- Set an appointment for the mediation session.

Our discussion now moves to the elements of the meeting.

CHAPTER FOUR

Beginning the Mediation

The mediator will bring the parties into the room and direct them to their seats. It may be necessary to attend to feelings or concerns that need immediate attention. When everyone is settled, the mediator will make an opening statement.

In the opening statement, the mediator lays the foundation for the operation of the session and begins to establish a rapport with the parties. When disputants choose Mediation, they often have an expectation that certain events will occur: the mediator will show appropriate interest in and concern toward them and their situation; the mediator will be competent to help them accomplish that which they have not accomplished on their own; a structure will be in place to safeguard their emotional and physical safety; and an established, orderly process will occur that, if followed, will move them from where they are in the dispute to where they need to be. Parties come to the session hoping these expectations will be satisfied. The mediator's opening statement should be carefully crafted to respond to those expectations.

Purposes of the Opening Statement

The purposes of the opening statement are to

- establish a safe environment to negotiate.
- establish the mediator's credibility and control of the proceedings.
- explain the mediation process and what will be asked of the parties.
- clarify the mediator's role.
- clarify procedures intrinsic to the process as well as housekeeping procedures.

- obtain necessary commitments from the parties concerning their involvement.
- establish the integrity of the process.
- address the stress often associated with conflict resolution.
- establish rapport, confidence, and trust with the parties to draw them in as quickly as possible.
- obtain clear commitments from the parties regarding behavioral guidelines. Parties want to know that their emotional and physical safety needs will be protected.
- be sensitive to concerns raised by the parties.
- answer questions.

The design of the statement should be supported by an effective delivery. As in all areas of Mediation, each mediator develops a style that is comfortable and effective. The styles vary in the degrees of formality, involvement, control, use of humor, and other areas. Regardless of the style selected, the delivery of the opening statement will make either a positive or negative impression that can have a profound effect on the process. The parties will quickly assess whether the mediator is someone who will be interested, effective, and helpful.

Presentation of the Opening Statement

Regardless of the type of dispute, the opening statement should include at least the elements (Moore, 1986) discussed below. A sample using a female mediator is presented.

Introducing the Mediator

The mediator should introduce herself.

Mediator: Good day, I am Sharon Leviton. I have been asked to be your mediator.

Establishing the Credibility of the Mediator

The mediator should establish her credibility by giving some background information on her experience and her credentials as a mediator. A lengthy history of training and experience is not called for. The parties are anxious to get on with their own agenda. They merely want to know that the mediator is trained and is competent to be helpful to them.

Mediator: I have been a mediator for the Community Mediation Service and in private mediation practice for 12 years. I have taught a family mediation course at the law school for 3 years and been involved in writing and training in Mediation for 13 years.

Introducing the Participants

The mediator should ask the names and identify the roles of everyone in the room. If unclear about someone's presence, she should deal with the matter promptly. If concerned about the absence of a possible key person, she should question the impact the absence might have on the proceedings and the outcome.

As the introductions are being made, the mediator should acknowledge each person. She should build on the human contact so that the parties are recognized as individuals, not as cases. She should demonstrate her respect and interest by taking the time to recognize each person with a brief greeting.

Mediator: I'm glad that you are here, Mrs. Brown.

or

I look forward to working with you both, Mr. Brown . . . Mrs. Brown.

Disclaiming Bias or Partiality

The parties must be convinced that the mediator is capable of acting as a neutral party and has no interest in the outcome. The mediator does not expect to gain benefits or payments resulting from the outcome. The issue of neutrality often creates contentiousness among dispute resolution professionals. Can the mediator be totally neutral? Can the mediator really separate her needs, value system, and biases from those being presented by the parties?

Reality would indicate that the mediator comes to the session with her own personal history, value system, perceptions about the world. These qualities are intrinsic to an individual's humanness. The key is not whether the mediator has strong feelings or bias toward the parties or toward the positions being taken or toward the issues at stake. Rather, the key is that she be able to make the separation behaviorally between her needs and those of the parties. Throughout this book reference will be made to the value of mediator awareness. Nowhere in the process is the need for self-monitoring more critical. The mediator must remain an advocate for a fair process and never for a particular settlement.

In claiming neutrality toward the parties and the issues, a mediator should disclose any relationship she has with any of the disputants that might raise concerns that she can in fact remain impartial. If the disputants claim that they are uncomfortable with the mediator's relationship with any of the parties, or have reason to doubt the mediator's ability to be impartial, they should state their concerns and gain clarity. The mediator may need to be replaced if the parties' concerns are not satisfied.

Mediator: I would like to assure you that I have no preconceived biases toward a particular solution or toward one of you. I spoke with you both by phone on October 14 to set an appointment for the joint interview that took place last Tuesday. I've had no other contact with either of you. Do either of you know me in any other context? Do you have any reason to believe that I might not be able to act in a neutral manner in this case?

Mr. Jones: No.

Mediator: Do you, Mr. Smith?

Mr. Smith: No.

The mediator should request and obtain a clear answer to each of the questions before proceeding. The idea is to prevent problems by anticipating them.

Explaining the Mediator's Role and Authority

The mediator probably gave this information during the intake. However, the mediator should not assume that the parties remember the explanation or that they heard it in the same way. This is an opportunity to demystify the process and to clarify misinterpretations or misunderstandings. Keeping the wording simple, the mediator should outline what the parties will do, what the mediator will do to assist the parties, and the scope of her authority.

Although the mediator might choose to use notes or a checklist during the explanation, she should observe the parties as she speaks to them. She must be aware of whether they are listening and understanding. This phase is one of the most important parts of the session. The mediator must lay the foundation for everything that follows. By taking the time to carefully explain, to answer questions, and to clarify faulty assumptions, the mediator proactively enhances the possibility of a productive session. She must protect herself from later charges that the parties were not fully informed concerning the process, the procedure, or their responsibility.

Mediator: My role as your mediator is to help you clarify the issues that you want to address, clarify the interests that you need to have met, keep you focused, help you work through impasses, and encourage you to keep trying when you get stuck. I do not have the power or the authority to make a decision for you. I will keep the process going, but you will control the content to be discussed. I have no authority to dictate the terms of the settlement. If you reach an agreement, I will encourage you to write a memorandum of understanding that summarizes the points. You may want to have an attorney review the agreement to make sure it is legally sound, clear, and all inclusive. The attorney can draft the agreement and put it in the form of a contract. If you do not reach a settlement, you are, of course, free to use some other forum. You do not lose your rights to go to court.

Emphasizing the Informal and Consensual Nature of the Process

As stated earlier, the mediator must maintain control of the Mediation from the introduction to the development of the memorandum of understanding. Maintenance of control requires the performance of four tasks. First, the parties must be kept on the agenda and prevented from long accounts of irrelevant events. Second, the mediator must balance her need for control with the clients' need to discuss the issues with each other during the session. Some mediators direct the parties to communicate with each other through the mediator. Some mediators encourage direct communication between the parties with minimal intrusion by the mediator. If too much control is exercised, the parties may feel suppressed and become disengaged. Not enough structure by the mediator can permit one party to gain control of the process. Third, the mediator must enforce the guidelines that have been established for the process, reminding the parties of their previous commitments to those guidelines. Fourth, the mediator must monitor her own feelings regarding the conflict to ensure that she is maintaining fairness. This structure is created to allow the process to occur. Within the structure there is wide latitude, unlike a court or other judgment-based setting.

Mediator: I will encourage direct communication between you. There will be no court reporters present.

Reviewing the Mediation Process

The mediator should outline the procedures that form the structure of the session.

Mediator: I would like to describe the process that we will follow. Each of you will have the opportunity to describe the situation as you see it. After the initial presentations, you have a chance to ask questions of each other The next step will be to identify the issues that you want to discuss and identify the needs or interests that you want to have satisfied. You will then begin the mutual problem solving.

Outlining the Procedure to Be Followed

Designating the Order of Initial Presentation
Designation of the order of the initial presentation can be determined in several ways. Often the mediator will ask the party who initiated the request for Mediation to present first. Under certain circumstances, the party most obviously in distress should begin. A party who is weaker might gain some internal psychological strength from being designated the first presenter.

Mediator: Mr. Jones, will you please begin?

Setting the Ground Rules
- *Interruptions will not be allowed:* A major stumbling block to joint problem solving is the failure of people to hear each other. The mediator should minimize interruptions by addressing the subject during the opening statement, seeking a verbal agreement not to interrupt, and monitoring compli-

ance with the agreement. The parties should be provided with a pad and pen to note observations or questions.

Mediator: Mr. Jones, will you agree not to interrupt Mr. Smith while he is speaking?

Mr. Jones: Yes.

Mediator: Mr. Smith, will you agree not to interrupt Mr. Jones while he is speaking?'

Mr. Smith: Yes.

Obtaining a clear verbal commitment will give the mediator leverage later when interruptions are likely to occur. The mediator can say, "Remember, you agreed not to interrupt." Eventually a mere hand gesture is often sufficient. Neither party wants to be seen as the agreement breaker. Breach of etiquette seems to occur more commonly in situations where the disputants have familiarity, a shared history, and proximity. The issues are not new, the arguments are no longer fresh, and the perceptions and the responses are predictable. The pattern of chaotic communication will be replicated in the mediation session unless the aware mediator prevents it from happening.

- *No personal attacks will be allowed:* The mediator must not permit rebuttals that are designed to attack the person. Allowing personal attacks to continue will ultimately breach the perception of safety.

- *Mediator will not be subpoenaed:* If the mediation case is not settled and the parties go to court, the mediator will not be subpoenaed.

- *Discussions are to be informal:* There are no court reporters and no tape recorders.

Explaining the Caucus and When It Might Be Used

The mediator should discuss the use of the caucus in the opening statement to avoid surprising the parties when she asks to meet with them separately.

Mediator: At some point during the session, it might be necessary for each of you to meet with me separately. This private meeting is called a caucus. The caucus allows you time to generate alternatives or proposals, helps me clarify my understanding, and allows me to manage tension levels and explore options that might be more comfortably discussed in private. Either you or I can initiate a caucus. What is discussed in these separate meetings will be held by me to be confidential. I will not reveal what we have talked about with the other party unless you give me permission to do so.

More discussion on caucusing will be given in chapter 6.

Addressing Confidentiality and Privilege

Mediators should describe the limits of confidentiality as it is provided for in their state. Chapter 9 discusses confidentiality and privilege.

Clarifying the Parties' Ability and Authority to Resolve the Case

The mediator should ensure that those people who have the authority to settle are present. If the proper parties or their representatives are not present or available, the Mediation should be postponed or rescheduled.

Dealing with the Logistics

The mediator should now give pertinent information about logistics and locations, such as

breaks
telephones and restrooms
arrangement for meals in extended sessions

Clarifying Time Commitments

The mediator should ensure that the parties are prepared to remain the full time originally agreed to. Having one of the parties unexpectedly announce the need to leave to attend a prior commitment is not only disruptive but also raises concern about intent, seriousness of purpose, and accountability.

Mediator: I have arranged to be with you until 4:00 P.M. Are you able to remain until 4:00 P.M., Mr. Jones? Are you able to remain until 4:00 P.M., Mr. Smith?

Clarifying the Parties' Willingness to Mediate in Good Faith

The mediator should again make a brief comment reminding the parties that an agreement reached will be a result of their efforts and their total involvement. The mediator should obtain from all parties and their attorneys, if present, their commitment to use their best effort to reach a solution everyone can accept.

Answering Questions

The mediator should ask the parties whether they have questions about the procedures. Spending necessary time early to clarify any issues reduces the need to play catch up later on.

The opening statement is designed to make the parties and their attorneys, if present, as familiar and comfortable with the process as possible. The mediator must assist the participants in understanding what is about to take place. The parties will not invest themselves if they have doubts or fears about the process, their safety, the mediator's neutrality, or some other unidentified concern.

Getting a Commitment to Begin

The mediator should end the opening statement by getting a commitment from the parties to begin.

Mediator: If there are no further questions about process, procedure, or ground rules, I suggest that we move on to discuss your issues. Are you ready to begin?

Mr. Jones: Yes.

Mr. Smith: Yes.

CHAPTER FIVE

Ventilation, Information Gathering, Problem Solving, and Bargaining

The next stage of the mediation process consists of ventilation, information gathering, problem solving, and bargaining. Bargaining can only occur after ventilation, information gathering, and problem solving. The mediator should use the following approach:

1. Ask each disputant to state his or her perception of the conflict. Hear all evidence pertinent to the dispute. Collect any evidence relating to the dispute, such as written contracts, canceled checks, receipts, and reports.
2. Clarify issues.
3. Determine whether the parties agree on the credibility of the incidents and information.
4. If the parties disagree, direct them to identify the differences and encourage them to account for the disparities.
5. Clarify remaining differences and see whether the disputants can form a common understanding.
6. Ask disputants to determine what they want and what they need to resolve the conflict. Help the parties differentiate between wants and needs, because this distinction will be crucial in negotiating an agreement.
7. Listen actively to the disputants' issues and feelings as they are talking.

The following interaction between the mediator and Mr. Jones is used to illustrate how ventilation might occur:

Mediator: Mr. Jones, I will ask you to begin by describing the situation as you see it. Please focus on defining the problem at this time. We're not ready yet to propose specific solutions. When you have completed your opening statement, I will ask Mr. Smith to define the problem. After this exchange, I will ask each of you to give some background of the problem, begin developing the issues that you would like to include in the discussion, and state the needs that you want to have satisfied.

Mr. Jones: I would like to be very polite about all this, but it's really hard. I am so sick and tired and frustrated with the reviewing and reviewing of memos and letters and trying to get Smith's attention. It's not fair that I should carry all the burden of this partnership.

Because the immediate dispute is often a residual of past or ongoing conflicts, the disputants frequently bring with them feelings of anger, frustration, disappointment, and revenge. Before they can deal effectively with problem resolution, the parties must deal with these feelings. During the ventilation period, mediators should be aware that people ventilate on different levels. Mediators must be flexible and able to cope with uncertainty and changing conditions. They must be prepared to adjust to the disputants' actions. A strategy of caucusing may prove beneficial in drawing out the disputants' feelings. While working in the ventilation stage, mediators need to be alert to the verbal and nonverbal cues given by the disputants. They must listen actively to the disputants' issues and feelings and show concern for the feelings expressed. Mediators are not expected to agree or disagree with emotions expressed. Listening does not require agreement. Mediators must be empathetic and attentive. The ability to reflect back the disputants' issues and associated feelings allows ventilation to be of maximum benefit.

Although ventilation works as a release for disputants, it also sets the parameters of the dispute for the mediator. The ventilation and joint disclosures describe the conflict situation and increase problem solving.

Mediator: I hear your frustration, Mr. Jones. It sounds as though the two of you have been round and round on this matter for years. That's why you both are here. Let's see if the two of you with my help can cut through the confusion and reduce the stress. Will you tell me something about the partnership arrangement?

Mediator guidelines for using the disputants' opening are as follows:

1. Learn about the parties' interests and priorities.

> ***Mediator:*** You seem to be most interested in clarifying employee responsibilities. Is that correct?
>
> ***Mediator:*** As you were talking, I wrote down your concerns. Could you prioritize those concerns?

2. Determine whether the parties agree on the credibility of the incidents and information.
3. Clarify the differences and see whether the parties can form a common understanding.
4. Ask parties to determine what needs must be met to resolve the dispute.
5. Help establish priorities among the issues.
6. Formulate clear goals.
7. Arrange agenda to cover general issues first, specific issues last.
8. Attempt to settle simple issues. Build on success.
9. Note overlapping interests of the parties and point them out.
10. Note parties' underlying needs and hopes. These are at the core of the dispute. Having them addressed and met will be the core of a resolution.
11. Reinforce areas of overlapping interests.

Bargaining and Negotiation

Prior to the bargaining and negotiation stage of the session, the mediator took an active role in the process by gathering information and serving as a conduit for the parties. During this stage, the disputants must take an active role and directly communicate their needs and interests. This will create the psychological ownership that will make the final agreement work.

Certain tactics seem to increase the chances of resolving an impasse in the negotiations. These tactics include the following:

1. Developing trade-offs that are equal in weight. Special attention should be given to understanding the equality of the trade-offs.

> ***Mediator:*** If Mr. Jones visits the children on Monday and Wednesday nights, you can go to your accounting class without having to hire a sitter. That meets his

need to see the children more often and your need to reduce sitter expenses this semester.

2. Establishing superordinate goals. The disputants develop outcomes that are more important than the conflict itself. If the disputants realize that only through a combined effort can the goals of both parties be achieved, and that the goal is urgent and highly desired, then resolution becomes of utmost importance. For example, settling a neighborhood dispute to protect the safety of the children in the area, or maintaining open lines of communication after a divorce becomes final to plan effectively for the benefit of the children might be interpreted as a superordinate goal.

Mediator: What would be the benefit of your ending this ongoing argument?

Mrs. Brown: Our children could play together.

Mediator: How would that be of benefit to you?

Mrs. Brown: My child would have a playmate. She needs that.

Mrs. Green: My daughter could ride her bike to school because she would have a companion.

Mediator: How would that be of benefit to you?

Mrs. Green: I wouldn't have to walk with her. The kids could watch out for each other. I wouldn't worry so much.

Mediator: You said that you want minimal communication with Mr. Jenkins when the divorce is over. What benefit might there be in maintaining open communication with him?

Mrs. Jenkins: The only reason would be for the children.

Mediator: Tell me more about that.

Mrs. Jenkins: The kids love him and want to see him as much as possible.

Mediator: You love the kids, too. How would your keeping the communication open with their dad benefit the children?

Mrs. Jenkins: I could plan more easily for them to get together. I could keep him informed on their activities.

3. Creating a synthesis. The mediator determines that the values in the conflict are total opposites and through problem-solving techniques helps foster a third view.
4. Allowing a graceful retreat. The mediator helps parties in retreating without loss of face. Face-saving can be at the core of negotiation. It often means backing away without seeming to have backed down.
5. Identifying and suggesting sources not apparent to the disputants. Mediators assist parties by generating alternatives. They often ask "what if" questions.

Mediator: What if you _____?
If this occurs, how would you respond?
If you look at the offer from his side, what do you see?
Have you thought about _____?
What would happen if _____?

Mediators must open options for consideration. They must allow the disputants to weigh and evaluate each option presented. Pressuring the parties into making a quick decision is counterproductive to the process. Parties must be given time to explore the consequences of a decision if a viable resolution is to be reached.

6. Narrowing the gap by pointing out similarities that exist and minimizing the differences. Often reminding parties of their stated needs and their proximity to achieving these needs will be encouragement to narrowing the gap even further. Recap:
   • Point out similarities that exist.
   • Minimize the differences.
   • Remind parties of their stated needs.
   • Look at the proximity to achieving needs.
7. Reducing tension. As was mentioned earlier, the mediator's sensitivity to the emotions of the parties is vital in conflict management.
8. Recognizing that bottom lines are illusionary.

Mrs. Smith: I'll never accept less than $125,000.

Mediator: If Amalgamated Inc. offers you $124,995, would you go to trial?

9. Opening constructive communication channels between parties and encouraging direct communication between them.
10. Setting limits. Although mediators should not pressure the disputants into making careless or rash agreements, their use of time limits to facilitate bargaining can be effective if handled judiciously.
11. Encouraging the parties to use the information they have about the other party in a constructive way.

Mediator: Knowing what you know about your husband's values, what part of this offer will likely appeal to him?

12. Emphasizing the parties' control over the solution.

Mediator: Do you really feel a jury is better suited to solve this problem than you are?

Mediator: Do you really believe a judge is in a better position to plan your children's future than the two of you are?

13. Doing a cost analysis of the difference between the offer on the table and the cost of going to court.

Mediator: Let's determine the expense of time off from work, the odds of achieving favorable jury findings, and the potential range of the numbers on the damage questions.

14. Doing a cost analysis of continuing the conflict. Included in the costing are the following:
    • emotional cost of continued involvement
    • time delays
    • attorneys' fees and court costs
    • risk of further damage to the relationship
    • risk of damage to the children, if applicable
    • publicity

Certain conditions may have an impact on the mediator's ability to assist the parties in their developing a resolution. These are

- the number of issues.
- the type of issues. Concrete issues are generally easier to resolve. Abstract issues, such as esteem, honor, and face-saving, are difficult to deal with.
- the number of disputants. As the number of parties increases, the number of issues also increases.
- the amount of time available. Disputants often limit their time, which pushes them toward action that may not include a review of the consequences. Conversely, it may prevent them from spending the time needed to resolve the conflict.
- the value system in effect.
- the communication channels available.

Many strategies and considerations have an impact on the bargaining or negotiations between the parties. The discussion of additional techniques for breaking deadlocks will continue in chapter 6.

If the mediators have used all their skills and do not feel that the case is resolvable, they will explain to the parties that they are terminating the session. If the parties are prepared to move toward writing a memorandum of understanding, the mediator will encourage them to do so. The memorandum of understanding should include the following elements:

1. Who is to perform.
2. What specific performance is expected.
3. When must the performance take place.
4. How is the performance to be handled.
5. How much is expected.
6. What will result in case of a breach of any portion of the agreement.

CHAPTER SIX

Techniques for Empowering, Caucusing, and Breaking Deadlocks

The terms *empower, caucus,* and *deadlock* or *impasse* have been mentioned and illustrated in various sections of this book. Because of the impact that these terms may have on the dynamics of the negotiations, they need additional consideration. This chapter will offer techniques to help the mediator create the perception of a "level playing field" without breaching the ethical concerns relating to neutrality, prepare a caucus, and move through deadlocks.

Empowering the Parties

Successful mediation cannot occur between nonequals. Those who believe they were full participants in arriving at a settlement are more likely to perform according to the agreement reached. The perception of power is a major force in a Mediation. Although the focus of responsibility for overseeing the process rests with the mediator, the real exercise of power and control over the dispute rests with each of the parties. The dispute and its issues belong to the disputants, and what they do to resolve the situation is up to them. The mediator's power becomes effective in the willingness to place responsibility where it belongs—on the disputants. The mediator empowers the parties by refusing to accept responsibility for the dispute, for the people, or for the outcome. The mediator places the burden of the problem and the power for problem solving with those who can best do what needs to be done.

Inequity of power often exists between the parties to a dispute because of differences in

- experience
- knowledge
- skills
- finances

Examples of power inequities might be experienced in disputes between:

- landlord and tenant
- principal and teacher
- parent and child
- employer and employee
- highly verbal person and less verbal person
- high I.Q. and low I.Q.
- higher wage earner and lower wage earner

Power ownership rests in the individual's perception. Although the mediator cannot do anything directly to either party to make them more powerful, he or she can create an atmosphere that enhances and maximizes each party's perception of having equal status within the negotiations. Examples of the effects of such a proactive approach might be as follows:

Perception of having enough information about Mediation:

"I understand what I am supposed to do, so I'll be all right."

Perception of personal physical safety:

"This feels like a safe place where we can really work."

Perception of emotional safety:

"I won't be verbally abused here."
"I won't be bullied here."

Perception of comfort with the process:

"There is a structure in the process that I can count on."

Perception that the mediator can be trusted to enforce ground rules and protect the process:

"I will be given a chance to be heard here."
"The mediator will explain things that I might not understand."
"Finally, someone will keep us focused so we can both talk."
"The mediator seems competent."

Perception of the mediator's neutrality:

"What I say is important. The mediator won't take sides."

Perception of full access to pertinent information:

"If I can see all the facts, figures, and records, we'll have a level playing field."

Perception of equal status in the Mediation:

"Neither one of us can settle this alone."

Techniques for Equalizing and Stabilizing Power in Mediation
* Understand the place of power in Mediation.
* Refuse to accept responsibility for solving the disputants' problems.
* Place the burden of the problem on the disputants.
* Place the power for problem resolution on the disputants.
* Use the premediation field trip to pick up clues about power disparities.
* Use information from the intake and the field trip to determine furniture arrangement.
* Bring the disputant perceived to have less power into the room first.
* Seat the weaker party in what might be seen as a more commanding position at the table.
* Assign homework as appropriate to ensure that the disputants investigate all aspects of the issues to be discussed.
* Suggest that the disputants consult other professionals such as attorneys, accountants, and/or appraisers as needed. Stress the importance of objective criteria as a tool in negotiations.
* Refuse to mediate if all information requested is not produced.
* Prepare and deliver an informative opening statement so that all parties are briefed together.
* Exercise control to the extent necessary. Use the agreements discussed in the opening statement as leverage to ensure that neither party monopolizes the discussion at the expense of the other. The mediator should also monitor personal control needs. The parties will behaviorally and/or verbally withdraw if they feel that they are the victims of an overbearing mediator.
* Caucus as necessary.

> *Mediator:*   Mr. Jones, are you aware that you are using your position to bully Mr. Smith? How do you think this will play with the other employees? You might not want the bad press your company will receive.

> *Mediator:*   Mr. Smith, let's take a look at other ways of ap-

proaching this. Let's also explore a fall-back position should Mediation fail.

• Do a cost analysis with the parties.

Mediator: Mr. Jones, you keep chiding Mr. Smith that you can always go to court if this fails. How much will it cost you in time, attorneys' fees, and employee relations? How about the possibility that Mr. Smith will counter sue?

• Be an agent of reality.
• Remain alert to verbal and nonverbal clues.
• Slow down or speed up the process when needed.

Mediator: Mr. Jones, you seem to be backing away.

Mr. Jones I don't have any idea what he's talking about. I might as well leave because I can't follow all this. It's too much too fast.

Mediator: Mr. Brown, you really seem to be dragging your feet. Could you tell me what's going on?

Mr. Brown: What's going on is that I'm completely stuck. I cannot come up with any way to make him understand how much this means to my family.

Mediator: What would happen if you don't push as hard on an immediate payment and concentrate on the repair work? Play with that a little bit.

• Continue to reassert the need for problem ownership.
• Continue to equalize the interactive power relations among the disputants.

Using the Caucus as an Adjunct to Mediation

A caucus is a private discussion between the mediator and one of the parties. To plan for success, discuss the use of the caucus in the opening statement. Build in an expectation that a private meeting might occur. Emphasize that all parties will have an opportunity to meet with the mediator in caucus. This assurance maintains the impression of neutrality and fairness.

The goals of the caucus are to

- gain information.
- increase or decrease tension.
- replace negotiation tactics that are not working with a strategy that might work.

Either the mediator or the parties may initiate a caucus. Statements or proposals discussed in caucus will not be revealed to the other side without permission.

Mediator: I will not reveal what we have talked about to the other party unless you instruct me to do so.

Thirty Reasons to Caucus

1. To separate one party from the intimidating conduct of the other.
2. To interrupt a prolonged stalemate.
3. To allow intense emotions to be vented.

   Mediator: Let's take a few minutes to drain off some of those feelings, Mary.

4. To encourage candor.

   Mediator: I sense that there is something critical that you have not revealed. It may be important that you open up so we can get to the root of the conflict.

5. To clarify perceptions or misperceptions.

   Mediator: Correct me if I'm wrong. I understood that you plan to continue operating the company locally. You alluded several times to a move. I am confused.

6. To change unproductive or repetitive negative behavior.

   Mediator: Every time there appears to be movement toward an agreement you introduce something new and sabotage the progress.

7. To limit unhelpful, unproductive communication.

   Mediator: Continuing to bring up old business is not helpful to the current discussion.

8. To clarify the negotiations procedure used by a party.

   Mediator: What are you trying to accomplish by taking that approach?

9. To stop negative procedures used by the parties.

   Mediator: If you remain positional, you are going to remain stuck.

10. To design new procedures for negotiations.

    Mediator: Let's begin to look at mutual interests and work from there.

11. To get information that may help generate alternatives.

    Mediator: Tell me more about the estimates. Let's fill in the missing information.

12. To address affective issues.

    Mediator: You seem to be shutting down. Would you like to share what you're feeling?

13. To address substantive issues.

    Mediator: Let's talk about the how much, where, when, and how of that idea.

14. To act as a sounding board for a party.

    Mediator: Try on your proposal. When you say it out loud, how does it sound to you?

15. To conduct reality testing of a party's proposal.

Mediator: How close does that come to meeting your needs?

_____

Mediator: Knowing what you do about Mr. Jones' interests, does this proposal meet most of his interests?

_____

Mediator: It sounds to me as though you haven't budged from your position, John.

16. To gain a better understanding of the realities of the conflict.

Mediator: You insist on putting your mom in a nursing home. She acknowledges the need for care, but she absolutely refuses to enter a nursing home. Each of you is threatening to dissolve your relationship. What is really happening here?

17. To help parties differentiate between feelings and behavior.

Mediator: You clearly have strong negative feelings about Ms. Jackson and what she has done. My job is not to attempt to take away your feelings. It is to help you separate those feelings from your behavior so you can make progress toward a settlement.

18. To move from positions to interests.
19. To deflate extreme positions.
20. To make appeals to common principles or superordinate goals.

Mediator: If you each take a little less initially, but pool your resources, you will both increase your assets and your earning power.

21. To separate inventing from commitment.

Mediator: Let's create other possibilities. Explore some options without committing to them. Perhaps a commitment will then seem to be the natural outcome.

22. To prevent a party's premature concession or commitment in joint session.

> *Mediator:* Help me understand. Very little in this proposal addresses your stated needs. What has changed? (The concern here is in the future performance of the agreement.)

23. To determine whether an acceptable bargaining range has been established.
24. To attempt to generate options.
25. To remind a party of the consequences of not reaching an agreement.
26. To identify possible areas of trade-offs.
27. To identify alternative solutions.
28. To gain understanding of the parties, their agendas, and their negotiating styles.
29. To increase the parties' perception of their power within the mediation process and over the outcome of the dispute.
30. To allow for a graceful retreat.

Phrases for Use in Caucusing

Getting started in a caucus is sometimes difficult. Students in training classes often ask for a phrase, a question, or a statement that they can use as an opener. The following are questions, phrases, and statements that the mediator might use to facilitate the caucus:

What would you think of _____?
Have you thought what might happen if _____?
Knowing what you know about him/her, how would he/she respond to _____?
When you put yourself on the other side of the table, how does your proposal sound?
If you put yourself in his/her situation, how does that offer make you feel?
Help me understand _____.
Something does not feel right about this. What is really going on? or What is it that you are not saying? or What is really concerning/scaring/worrying/angering you? (These questions are attempts at tuning in to the person without becoming threatening. They can always back away, and you can apologize for misreading them.)
Let's take a look at _____.

This is what I am hearing. Is that what you meant to say?
Refer to what you stated as your need. Does this proposal address that need? Does it also address his/her need?
If you suggested _____, what would happen?
What did you hear him/her say about that issue? What do you understand about his/her need in that?
Could you elaborate on that?
Is there anything else?
How does that _____?
What do you want it for?
What if _____?
Are you aware that every time you _____, you _____?
Did you hear him/her refer to caring about that, also?

Mediator Guidelines for Preparing to Caucus

1. Time the caucus. Knowing when to initiate a caucus is critical to its success. There are three major concerns that if triggered individually or in combination indicate the need for a caucus. These concerns are

   - the need for information.
   - the need to affect tension levels.
   - the need to change negotiation approaches of either or both sides.

2. Announce the decision to caucus.

   Mediator: At the beginning of the session I explained that I might meet with each of you separately during the session. I would like to do that now.

3. Clarify room assignments for the caucus. The caucus room can be arranged in several possible ways.

   Assign each party a room. The mediator shuttles between the two rooms. Advantages are

   - the parties have fewer moves to make.
   - the parties keep paperwork in order.
   - the parties can relax and prepare while waiting their turn.
   - the mediator is saved from dealing with possible hostile exchanges.

Designate a caucus room and a waiting room. Parties are shuttled between the waiting room and the caucus room. The advantage is that the arrangement will work when limited facilities are available.

4. Clarify arrangements for the caucus.

   Explain location of the rooms.
   Determine order of the caucus.
   Explain location of telephone and snack areas.
   Determine amount of time to be spent in caucus. Explain that disparate amounts of time may be spent in caucus.
   Reaffirm commitment to meet with each side.

5. Determine the length of the caucus. The caucus should continue until the mediator feels that the purpose has been satisfied or the deadlock has been broken. The following are questions for the mediator to consider:

   Is progress being made?
   Do I have a clear understanding of the real issue?
   Did I uncover the real agenda?
   Was anything new revealed that I need to storehouse for future use?
   Did the party evidence trust in me?
   Are the tensions being managed?
   Did I succeed in generating options?
   Is this going anywhere?

   If the caucus fails to produce movement, the mediator should terminate the caucus. If the caucus is productive, the mediator should clarify what information is to be revealed to the other party.

   Mediator: Is there something that you would like me to take back to the other party?

   Martha: Yes, you can tell them that I would be willing to

   _____.

   Mediator: Do you want me to say anything about

   _____.

   Martha: No, not at this time.

   Mediator: Let me be clear. I am to discuss your willingness to buy lots 4 and 5, but not mention lot 10.

   Martha: That is correct.

6. Leave an assignment for the waiting party. Before moving to caucus with the other party, leave a question or information for the party to consider.

Mediator: While I'm visiting with Mr. Brown, think about

_____.

After the first round of caucusing, the mediator will decide whether to

• remain in caucus.
• reconvene the joint session to continue negotiations.
• reconvene the joint session to terminate the Mediation.

Breaking Deadlocks

When a mediation session has reached an impasse, the mediator must attempt to work through it.

Reasons for Impasse
• Key decision maker is not involved or not engaged. The person is physically not present, is present but not participating, or is shifting the decision making to others.
• Goals of the parties have not been clearly established.
• Benefits of settlement have not been clearly articulated and weighed.
• Costs of maintaining the conflict have not been clearly assessed by the parties.
• Each side is locked into the need to punish the opposing side.
• One or both parties see a benefit to maintaining the conflict.
• Fear or concern of losing face exists.
• One or both sides believe that another forum will resolve the dispute to their advantage.

"My lawyer told me I would win in court."

• Legitimate interests of all parties have not been adequately addressed by the settlement proposal.
• Hidden agendas of parties have not been discovered by the mediator.
• Emotions have replaced reason.
• Parties have stopped listening to each other. They may be rehashing the same familiar themes without adding any fresh ideas, using the same tactics that historically have been inef-

fective, and/or focusing on personal attacks rather than on issues.

Techniques for Getting Movement to Occur

EXPLORE alternatives.

SEEK trade-offs.

REFER back to needs and priorities stated in each party's opening statement.

REASSESS parties' needs.

REASSESS parties' priorities.

POINT OUT movement that has taken place.

Selectively USE information gathered from the caucuses.

HELP the parties phrase proposals in the form of hypothetical questions. Encourage them to "try on" proposals.

SERVE as an agent of reality.

- Have each party think through his or her demands. They should then question the rationale for the demand.
- Focus each party on the other's arguments.
- Ensure that each side understands what the other party is saying.
- Make sure that each side knows what the other side wants included in the agenda.
- Check out the validity of information.
- Probe how each party might act to resolve the conflict.

DEFLATE unrealistic demands.

NARROW differences between parties on the issues.

ENCOURAGE the parties.

"You're making progress. Don't give up now."

USE external pressures such as deadlines, press, and/or public reactions.

EMPHASIZE the parties' control over the solution.

Mediator: Do you feel that a court is better suited to resolve your business problems than you are?

DO a cost analysis of the offer.

Mediator: If you accept what's on the table, you'll have close to what you stated your needs to be. Compare that to projected expenses, time off from work, and the odds of achieving a favorable jury finding.

CREATE the idea of a new possibility.

> *Mediator:* You can take this offer and use it for something tangible, such as health care, a car, or college tuition.

HELP parties save face.
ESTABLISH superordinate goals.
SET time elements in perspective.

> *Mediator:* Even if the court awarded you the full amount, how long will it take until you have the money in your possession?

CHALLENGE the party.

> *Mediator:* Mr. Jones has made several concessions. I am concerned that unless you show some flexibility you might not reach a settlement.

LOOK for solutions that are low cost to one party but of high value to the other party. Examples might be an apology, a quick payment, or assistance of some kind.
BRAINSTORM with the parties to get options on the table.
APPEAL to fairness.
SEEK and utilize objective standards.

Potential Outcomes of Not Reaching an Agreement

The following are questions the mediator might ask in exploring potential outcomes of not reaching an agreement:

Do you think you can win in court?
How certain are you? Scale the possibilities from one to ten.
What risks are you willing to take?
What if you lose?
What will you lose?
What will your life be like then?
What impact do you think a victory in court will have on your ongoing relationship with the other party?
Who else besides the primary party and you might be affected?
If you were in the other party's place and this proposal were made to you, would you accept?

Let's explore the

1. strengths of the case.
2. weaknesses of the case. What would you like to be different about your case?

On a scale from one to ten, rank the following:

1. This settlement approaches my best day in court.
2. This settlement mirrors my worst nightmare in court.
3. This settlement is amenable to the other party.

Finally, in attempting to break impasses, recognize the importance of the mediator's skills and personal attributes:

- Continue to build trust by being trustworthy.
- Ask questions.
- Make observations.
- Raise risks.
- Remain neutral.
- Be optimistic.
- Be realistic.
- Be inventive.
- Be sensitive.
- Listen for understanding.
- Be sensitive to disputants' physical needs.
- Have a reason for what you ask and say.
- Be clear about the limits of your role.
- Protect the integrity of the process.
- Plan and be prepared.
- Be aware.
- Allow the dispute to be theirs.
- Allow the settlement to be theirs.

CHAPTER SEVEN

Family Conflicts

Family Mediation is sometimes seen as being limited to divorce and separation cases. The reality is that the mediation process can be used in all types of disputes between family members. Mediation provides for neutral assistance in settling disputes arising from almost any area of the life cycle.

Those who are committed to finding a way to solve a problem regardless of its nature can effectively accomplish a resolution. The neutral third party provides structure for the discussion and assists the parties in seeking a mutually acceptable settlement of the dispute.

Examples of Mediation Along the Family Life Cycle

- Premarriage Mediation—the development of marriage agreements prior to the ceremony.
- Disabled and handicapped—conflict between a disabled family member and the family caregiver or between family members concerning responsibility for the disabled member.
- Teenager and parent—conflict in multiple areas such as:
  Child's curfew
  Child's choice of friends
  Child's social lifestyle
  Child's household chores
  Child's school performance
  Child's attitude
  Child's allowance
  Parent's method of discipline
  Child's problems with siblings
  Child's use of the family car, phone, or other family resources

Child's use of drugs/alcohol/tobacco
Parent's nagging, lecturing, or invasiveness
Mutual problems in communication

Mediation can be a forum for resolving immediate disputes and setting up some ground rules for parents and children to live together. Each gains a voice in establishing the agreement, and each has prescribed responsibilities for compliance. The mediator will probably have the parties come to some common understanding of "fairness" and "power" as they discuss the issues of the dispute.

- Relocation—conflicts surfacing from a family move. Issues surrounding dual careers and care of dependents are often the basis of the dispute.

> *Mary:* My job is as important to me as yours is to you.
>
> *Jim:* I've spent years trying to get to this place in my career. And now we have the opportunity and you don't want to go. We may never have this chance again.
>
> *Mary:* But I was promised a promotion if I stay with the company two years. I would have to start all over again. The kids are settled and our parents are here.

- Teenage pregnancy—conflict between unmarried pregnant teenagers and parents.
- Homosexual relationship—similar conflicts as exist in heterosexual relationships. Mediation has the power to enable people to meet their needs when the law does not exist to provide a remedy.
- Unmarried cohabitants.
- Pre- and postretirement Mediation—conflicts concerning relocation, time allocation, and finances.

> *Jane:* He's going to retire next month, and I'm really concerned.
>
> *Mediator:* What's your concern, Jane?
>
> *Jane:* He's planning absolutely nothing. He has no hobbies. His life is his work. Now he's just going to hang around the house and expect me to hang around with him. I need some agreements about this.

- The elderly and adult children—conflicts concerning institutional versus home care; conflicts between siblings about shared responsibility for elderly parents; conflicts about visits, health care, activities, and unresolved concerns.
- Return of adult children—conflicts between parents and adult children who have returned home temporarily or permanently. The issues are often about shared housekeeping responsibilities, household resources, grandchild care, space, and privacy.
- Remarried families with children.
- Domestic violence.
- Wills and estate planning.
- Separation and divorce—conflict about property settlements and support payments.
- Custody and visitation.

Mediation of the above situations should not be confused with psychotherapy or counseling, which may be needed in these same situations. Family or individual therapy may be required in such cases to lay the groundwork for the practical process of Mediation.

The procedures outlined in earlier chapters are used in family mediations. Family disputes have some added dimensions that are not always as striking in other kinds of conflicts, particularly those in which the parties have a remote or limited relationship.

Twelve Guidelines for Use in Family Mediations

1. Understand what makes family conflicts unique. Special considerations concerning family conflicts that affect family Mediation include:
   - a shared history involving family, friends, homes, businesses, activities, hopes, and dreams.
   - the emotional baggage that disputants bring.
   - the need to maintain a relationship because of children, a business partnership, property, certain affiliations, and/or family.
   - familiarity.
   - long-standing unresolved issues.
   - predictable behavior patterns.
   - a predictable pattern of communication.

- shared personal relationships.
- shared family property.

2. Recognize behavior. The parties' behavior in the mediation session will tend to mirror the way they interact with each other outside of the session. The effective mediator will be aware of the parties' behavior and what is being communicated both verbally and nonverbally.

Behaviors that enhance the possibility of settlement include:

- an openness to ideas and options.
- a concern about mutual interests of the parties.
- an interest in exploring possibilities.
- an involvement in problem solving.
- a willingness to be responsible and take responsibility for reaching a resolution.

Behaviors that diminish the probability of settlement include:

- withdrawing from the discussion.
- replaying familiar tapes and offering nothing new or current.
- pushing "buttons" that always draw negative responses or add to conflict.
- threatening to leave the session while negotiations are in progress.
- giving in but not resolving anything.
- sulking, pouting, and becoming passive rather than generating ideas.
- being verbally abusive.
- failing to take responsibility by coming late or missing sessions or skipping assignments.
- sabotaging the progress by "foot dragging."
- failing to stay focused on the discussion.
- remaining positional.
- continually bringing up past actions, and hurts.
- blaming and accusing.
- withholding pertinent information.
- seeking control over the other party, the mediator, and the process.
- operating from a victim position.
- placing the locus of responsibility elsewhere.

3. Recognize feelings. Read chapter 11 for a list of the stressors that parties often experience in connection with a Mediation. Recognize the need to acknowledge the parties' feelings. Use information in chapter 11 as a guide for listening and responding. Typical feelings include

hurt	resentment
fear	impotence
abandonment	revenge
failure	disappointment
grief	despair
alienation	relief
anger	denial
loss	guilt
embarrassment	ambivalence
helplessness	free-floating anxiety
hopelessness	betrayal
frustration	

4. Recognize power. The alert mediator will soon recognize the power plays and control tactics that are brought into the Mediation by one or both parties. What normally occurs outside of the session will be replicated in the session unless the mediator intervenes. Power disparities often exist between spouses. They almost always exist between parent(s) and minor children. Read chapter 6 for information on empowering.

The following vignettes illustrate issues of behavior, feelings, and power.

"Not with the Other Woman"

John and Judy married 15 years ago and have two children—Jill, age 6, and Jack, age 4. Judy is a doctor sharing an OB/GYN practice with two other women. John is a radiologist in solo practice. They have lived apart for 2 months. The court has referred them to Mediation because of their conflict over John's access to the children.

How Judy Sees the Conflict
John left the children.
The children are unhappy about seeing their father.
When the children come home after a visit with John, they cry and are upset for an hour or more.

The children are devastated by the separation.
John has destroyed their plans.
John gave her no reason for leaving.
John is exposing the children to some other woman.

How John Sees the Conflict
He left because he was spending most of his spare time in the evenings and on weekends taking care of the kids at home while Judy was at the hospital, at a meeting, or sleeping because of her irregular hours.
He did not leave because of the other woman.
The girlfriend has not spent the night when the children were present but has met them.
Judy wants to control his life by complaining about where he lives, the new Mercedes he drives, and who he sees.

What Judy Sees as the Solution
John should come back home.
The mediator should help her get him back.
If John isn't coming home, the girlfriend should not be around when John has the children.
Visitation should be limited to two weekends a month.

What John Sees as the Solution
The mediator should straighten Judy out and make her accept the situation.
The children and the nanny should spend half the time with him on a weekend or two-week basis.

Underlying Issues

JOHN	JUDY
Feelings	*Feelings*
abandoned	abandoned
annoyed	betrayed
angry	bitter
competitive	cheated
controlled	determined
imposed on	outraged
frustrated	jealous
	vindictive
	threatened

JOHN	JUDY
Power	*Power*
control	keep visitation minimal
	use children if John
	does not return
Behavior	*Behavior*
closed down	looks down at floor
gazes at wall	looks away from John
turns away from Judy	interrupts when John
looks to mediator to	speaks
solve problem	looks to mediator to
	solve problem

"You're a Hundred Miles Away"

Don and Kim have been married for 15 years. They have a daughter, Lynne, 12 years old, who was diagnosed 5 years ago as a diabetic. They separated 6 months ago after Kim filed for divorce. All the details of an Agreed Decree have been worked out but not yet formalized into the final document. An important event has taken place, throwing the previous agreement into disarray. Kim has primary possession under a joint custody arrangement. She is an executive vice-president of a bank and has been offered a promotion as a bank president in Boston. She wants to take the job. Don, who thought he had an agreement, has just been told this and is very angry. He does not want Kim to leave Biloxi, because it means Lynne will move with her mother. Their lawyers thought it would be useful to mediate just this issue to save the rest of the package.

How Kim Sees the Conflict
Don's demand for access every weekend is ridiculous and
  unworkable now.
She has always been Lynne's caregiver.
Don has been uninvolved in the health issue.
She supervises the diet and the periodic blood testing.
  Lynne needs to be with her.

How Don Sees the Conflict
He is worried about the disruption in Lynne's life.
He doesn't know how they will deal with the transportation. The expense is going to kill them.

The agreement was that he would have access to Lynne every weekend and overnight when she wanted to stay.

What Kim Sees as the Solution
She feels no responsibility for the problem.
She didn't seek the job; it just happened.
She has no power to help its resolution.
The solution is up to Don. He has to change.

What Don Sees as the Solution
Kim is obligated to stay in Biloxi. That was the agreement.

Underlying Issues

DON	KIM
Feelings	*Feelings*
concerned	self-righteous
betrayed	excited
outraged	eager
sad	
Power	*Power*
will use anger to attempt to gain control	will use Lynne's medical condition to gain power
Behavior	*Behavior*
positional	positional
hostile	emotionally and behaviorally distant
	Unannounced Issue: fear that if she moves and Lynne visits her dad and former peers in Biloxi, she won't want to return to Boston

5. Be alert during intake. Read chapter 1 for a list of concerns during intake. Be clear about the parties' expectations of Mediation.
   - Is their purpose to save or dissolve the marriage?
   - Do they see Mediation as a forum to beat up on each other?
   - Are they wanting to circumvent disclosure?
   - Do they want the mediator to take responsibility for them?

- Do they use the mediator as an advocate for their position?
- Do they use the children inappropriately?

Be clear about the parties' emotional capacity to be responsible and about their mental capacity to participate.

6. Set ground rules carefully and sensitively. Parties often require the structure that the mediation process represents. The purpose of setting ground rules in Mediation is to allow the process to happen. If the parties were willing and able to establish and abide by their own ground rules outside of the process, they would not need a facilitator. By establishing and enforcing rules of decorum, rules of procedure, and expectations of performance, the mediator provides a reliable structure within which the parties can work. In family situations where an ongoing relationship exists, the mediator becomes an educator. The problem-solving skills that the parties learn can be applied in their future negotiations.

7. Be sensitive to continued impasses, missed appointments, or other delays. Question the implications of these behaviors from the standpoints of

- fairness to the other party.
- the potential for having to abort the Mediation.
- the cost and consequences of impasses.
- the unexpressed needs met by the delays.
- the appropriateness of the case for Mediation.

Mediators may need to caucus with the parties to state their concerns and explore the root of the behavior.

8. Know when to include the involved children in the Mediation. Considerations include

- age of child/children.
- intellectual and emotional capacity of the children.
- intention for including the children.
- desire of the children to participate.
- verbal ability of the children.
- maturity of the children.
- amount of conflict between the parents.
- use of the children as a pawn in the conflict.

Do a cost/benefit analysis of including the children.

Will their participation be harmful to them?

Will their participation be helpful to them?
How will they be prepared for participating?
How will their input be used?
How extensive will their involvement be?

Their level of involvement should be examined in premediation interviews with the adult parties and discussed with the children separately.

Is it necessary to have a guardian *ad litem* appointed? Would it be helpful to have the guardian *ad litem* at the Mediation? If so, what role would he or she take? What is the guardian's perception of the situation and his or her assessment? Use the guardian *ad litem* as a resource and helper if possible.

9. Use homework assignments to

- empower parties.
- bring information to the table.
- discover how willing the parties are to be involved.
- dispel myths concerning family income and expenditures.
- identify differences in perception about information and issues. It is often the perception rather than the facts that are at the base of a conflict.
- meet with a lawyer, an accountant, and/or an appraiser for information necessary in the negotiations.
- create a cooperative, problem-solving mode.
- educate the parties in problem-solving techniques for use in future negotiations.

As with every other aspect of the mediation process, having a plan with a clear purpose is a key element. What is the assignment, what is its purpose, when is it to be accomplished, and what if one or both parties don't comply?

10. Be creative in generating options, especially when children are involved:

- Explore possibilities.
- Brainstorm.
- "Try on" ideas.

A major benefit of Mediation is the opportunity to be creative.

Mr. Jones: I want visitation with the children more than two weekends a month. Can't she understand how important my children are to me?

Mediator: Seeing your children means a great deal to you.

Mr. Jones: How can she do that to me! I've raised those kids. I've carpooled every day, paid for their schooling, cuddled them when they cried or were sick. I'm the one they come to when there's a problem. My little boy needs to have me around.

Mediator: It sounds as though you need him, too.

Mrs. Jones: You should have thought of that when you decided to leave us .

Mr. Jones: I wasn't leaving the children. I'm a good father.

Mrs. Jones: You'll get what the court would give you. That's all I have to do.

Mediator: The children are important to you aren't they, Mrs. Jones?

Mrs. Jones: Of course they are. I am a great mother.

Mediator: I am sure you are. What do you think are necessary ingredients in child raising?

Mrs. Jones: Love, caring, education, safety, playing, morals.

Mediator: Mr. Jones, what do you think are necessary ingredients in child raising?

Mr. Jones: A lot of love, parents who care, a good school, having fun, having friends, values.

Mediator: It sounds as though you both want the same things for your children.

Mrs. Jones: But he left his kids. He does not deserve to have more time.

Mediator: How about the kids. Do they deserve to have more time with a dad they apparently love very much?

Mrs. Jones: Well . . . . When do you want to see them?

Mr. Jones: I could babysit the two evenings a week that you work. You wouldn't have to worry about a sitter. I can also continue to do John's morning car pool.

Mrs. Jones: No. Jim, I don't want to have to deal with you all that much. Instead of the morning car pool, how about Saturday morning soccer? That would give me some free time.

11. Be caring and empathetic and stay in your role as mediator.
12. Do not let personal issues contaminate the process. Sometimes mediators may have cases that too closely parallel events either past or present in their own lives. The issues, the personalities of the parties, and/or the communication style of the parties may trigger a panoply of feelings. Feelings and issues thought to have been resolved suddenly resurface. The experience reflects the mediator's own humanness. The key to remaining effective is to allow the feelings to be what they are. Mediators can't change feelings. They can monitor their behavior so as not to use the Mediation to resolve the personal conflict that they might be experiencing.

CHAPTER EIGHT

Co-Mediation

The authors of this book co-mediate 98 percent of their cases. Although some use co-mediation primarily for training novice mediators, we recommend its use as the norm rather than the exception. A mediator fills many roles and performs many tasks. Having an effective co-mediator can reduce the pressure on the lead mediator and allow him or her to better serve the process and the parties. The co-mediator provides

- an extra pair of eyes.
- an extra pair of ears.
- validation and support to the lead mediator.
- the monitoring of the progress of the Mediation.
- support to the process.
- the ability to fill in as necessary for the lead mediator.
- assistance in the caucus sessions.
- feedback to the lead mediator.
- observations concerning the parties:
  What are the parties not saying?
  What is pushing their hot buttons?
  What is their stated agenda?
  What is the hidden agenda?
  What is the posturing about?
  What is drawing the parties together?
  What ideas or words peak their interest?
  When is the mediator losing them?
  What is really important to the parties?
- help in determining times to break and caucus.

What follows are some of the questions often asked about the mechanics of co-mediation.

- When is co-mediation particularly useful?

  All cases

  Complex cases

  Multiparty cases

  High-stress, potentially hostile cases

- Do co-mediators mediate equally?

  No. One mediator initially takes the lead position and remains in the lead. If a change is to be made, it should occur only after consultation and agreement between the mediators or on pre-arranged signals established between the mediators.

- What considerations are used in selecting the lead mediator for a particular case?

  Nature of the case

  Strengths of the mediator

  Specialized skills of the mediator

  Personality of the mediator

  Profile of the clients

  Ages of the clients

  Experience of the mediator

  Culture of the clients

  "I led last time; it's your turn."

  Illustrations of the Above

  1. In certain cultures, a female mediator will be afforded respect but not be taken seriously. A male mediator might therefore be selected as the lead.
  2. High-risk, potentially hostile cases would call for a mediator with strong crisis-management skills.
  3. A case involving an angry husband, a passive wife, and male attorneys representing both sides might be well served by a strong, calm, female lead mediator.
  4. In a dispute between two elderly parties, selecting a lead mediator perceived to possess greater life experience might be necessary.
  5. In a cross-cultural case, selecting a mediator who is knowledgeable about both cultures is important. Understanding the spoken language is not enough. Being sensitive to their cultural values and cultural perspectives is essential.

- Is there an additional charge for co-mediation?

No. These authors feel that the clients are best served by co-mediation. Therefore, no additional charges are made for the use of two mediators. In addition, the authors believe that the use of two mediators is not an optional matter.

- How is the co-mediator introduced?

The lead mediator introduces the co-mediator during the introduction phase of the process. The co-mediator might give a brief statement of credentials. This would be a good time to emphasize that the function of the co-mediator is to enhance the process, not to be a representative of either party.

Mediator Guidelines for Effective Co-Mediation

1. Give an impression of structure and cooperation.
2. Clarify roles.
3. Check out your ego. Competition, hidden agendas, and power plays among mediators will sabotage the Mediation.
4. Orchestrate changes in mediators carefully so as not to cause confusion.
5. Avoid interrupting the other mediator.
6. Avoid interrupting transactions between the other mediator and a party.
7. Avoid introducing a competing idea during a transaction.
8. Avoid introducing new material prematurely.
9. Signal the other mediator if his or her approach is obviously not helpful.
10. Have clearly defined eye signals or hand signals concerning times to caucus or to switch leadership. The more experience that co-mediators have with each other, the easier will be the development, utilization, and recognition of these signals.
11. Caucus with the co-mediator as appropriate to share information and to develop strategy. Keep the caucus brief. A good time for co-mediators to caucus with each other may be during times of mediator movement between caucus sessions.
12. Learn to be comfortable with silence. If the lead mediator is successfully directing the procedure, co-mediators must support those efforts. Do not prolong the procedure or intrude inappropriately out of a concern that silence will be seen as impotence or a lack of interest on the part of the co-mediator.

13. Avoid becoming trapped into alliances with either party.
14. Do not disagree with the other mediator in the presence of the parties. Do this privately.
15. Observe the verbal and nonverbal behavior of the parties. Make note of phrases, facial expressions, comments, and reactions that might have importance. Be prepared to share this information with the lead mediator.
16. All participating in the co-mediation process should be trained mediators regardless of additional or primary professional training. An observer to the mediation process, either student or colleague, should not be confused with someone serving as a co-mediator. An observer has no role in the mediation process except that which may be attributed to him or her by the mediator.
17. Remember, co-mediation does not mean "equal mediation."

When there is preplanning and a clear understanding of procedure, a delineation of tasks, and a willingness to support team efforts, the opportunity for an effective co-mediation is maximized. Failure to preplan and to follow through could result in chaos.

CHAPTER NINE

Ethical Considerations in Mediation

Ethics are guidelines or principles of conduct that govern an individual or group and are based on the morals and values of our culture. The mediator is guided by a core set of values in making decisions concerning right or wrong in a given situation. When confronted with ethical dilemmas, the mediator should consider: Will my action conform to or violate the legal standards adopted by our society? Will my action conform to or violate my internalized standards of right or wrong? Will my action be defensible when scrutinized by the public or other members of my profession?

Elements of Mediation is about delineating, defining, and presenting the qualities and the procedures which form the ethical core of the mediation profession. The consumer (client) expects and has the right to demand the mediator be trustworthy, reliable, honest, fair, impartial, consistent, credible, competent, concerned. The mediator will explain the process and be the gatekeeper to ensure the client and case are appropriate for the process. The duty of the mediator is to present an informative and instructive opening statement that describes procedures, roles, and responsibilities.

Clients will want to know the costs and method the mediator will use to help them move from their current impasse to a resolution of the dispute. They expect the process to be confidential, emotionally and physically safe, and that the agreements agreed upon will be legally and financially sound and in the best interest of all involved. Clients, therefore, must be held accountable by the mediator to obtain legal, financial, and other professional advice in order to make informed decisions during negotiations. The following are nonexhaustive guidelines to foster the goal of the profession:

1. Reveal conflicts of interest. The mediator must reveal any business or personal interest in or any benefit that might accrue from the outcome of the Mediation.
2. Reveal previous relationships with any of the parties.

3. Charge a fair fee based on current standards.
4. Communicate confidentiality and privilege. Confidentiality refers specifically to what goes on in the caucus. Mediators will not reveal in open session any information learned in caucus without permission of the party. Confidentiality can only be asserted by the mediator on behalf of the party not on behalf of the mediator. There are some exceptions to confidentiality. Each mediator should be aware of state and federal laws affecting this issue. Such exceptions may include child abuse, homicide, suicide, and commission of a felony. Privilege refers to the joint session. No information revealed in joint session can be used in court against the other party.
5. Communicate the rules of Mediation. Give a clear explanation of the mediation process and the procedures involved. An outline of the opening statement is given in chapter 4.
6. Advise that the mediator is not a witness for either side.
7. Limit contacts with parties to scheduled sessions. Reveal any contacts between sessions to all parties.
8. No subpoena will be served immediately before, during, or immediately after a mediation session.
9. Recognize the importance of problem ownership. Both the problem and the resolution belong to the disputants. Do not exceed the mediator's responsibility or function.
10. Remain impartial to all parties. If the mediator or the parties find that the mediator's impartiality has been compromised, the mediator should offer to withdraw from that Mediation. The mediator's commitment is to aid all parties in reaching a settlement.
11. Encourage full disclosure of information.
12. Do not give legal or other professional advice. As appropriate, encourage the parties to seek legal, financial, or other professional advice before, during, or after the mediation process.
13. Avoid giving personal opinions.
14. Postpone, recess, or terminate the Mediation if it is apparent that the case is not appropriate for Mediation or the parties are unwilling or unable to participate effectively in the mediation process. Refer to chapter 1.

CHAPTER TEN

Dealing with Hostility

Participants in Mediation often experience heightened emotions. For the mediator to enter a session without recognizing the possibilities for both expressed and unexpressed hostility is to enter unprepared. As a result, unrealistic expectations may be maintained. Mediator awareness, ground rules, empathy, clearly stated expectations of behavior, caucusing, crisis-management skills, and control by the mediator are key elements to effectively diffusing and utilizing hostility.

The mediator is responsible for assessing the levels of feelings that exist. The mediator must deal with them in ways that will enhance the Mediation attempt and prevent nonproductive escalation. In this way, the mediator may protect those who may be inadvertent victims of unprovoked hostile actions or expressions.

Guidelines for Handling Hostility and Hostile Gestures
1. Deal with hostility at its first expression.
2. Handle physical violence by immediate separation of the parties.
3. Learn to attend to your own senses or "inklings" that forecast impending crisis in a mediation session.
4. Remain in firm control of the mediation session at all times.
5. Redirect the expressions of strong feelings, if at all possible. Remember, not all expressions of hostile feelings are counterproductive.
6. Pay attention to the physical arrangement of the mediation room to provide the best alternatives for handling hostility.
7. Enforce all ground rules established in the mediator's opening statement.
8. Caucus, as needed, to allow for ventilation of hostile feelings.
9. Do not attempt to argue with the disputants about the cor-

rectness of their feelings. The anger and hostility are real to them. Often a disputant merely wants to be understood and have his or her feelings given credibility. All expressions of feelings must be heard, acknowledged, and understood as much as possible. They should be given credibility and used constructively.

10. Recognize that hostility can be a sign of impending crisis in the life of the disputant.

11. Prevent guns, knives, and other weapons from being brought into the mediation session. Remove objects that could become lethal weapons in the hands of a highly agitated person. Ordinary office accessories such as letter openers, scissors, bookends, and small electronic equipment all have the potential for injury if not lethality.

12. Provide for security of mediation offices and agencies, and establish standard operation procedures with law enforcement and paramedical agencies.

13. Consider a "buddy system" for mediators to enable them to respond to and to assist each other when needed. This is especially important whenever Mediation takes place after usual business hours.

14. Terminate the mediation session at the point when it becomes clear that progress is no longer possible because of potential hostile actions. This will salvage the possibility of returning to Mediation at a later time.

15. Be prepared with skills, protective measures, knowledge, and backup support for every Mediation.

Special note: Read the chapters that discuss crisis management and safety.

CHAPTER ELEVEN

Stressors, Stress Management, and Crisis Intervention

The potential for client stress and tension may be at maximum levels before, during, or immediately following a session. At these times, the client may experience extreme feelings of fear, anger, grief, hostility, and/or alienation from his or her self-concept, family, and society. Mediators must recognize the potential for high stress and the crises that may result. They must be able to prevent, wherever possible, the acceleration of tension through a proactive approach and to manage any crisis that should occur. Prevention, preparation, and effective intervention can provide for the increased safety of all parties to the Mediation or attendant to the process. Mediation cannot continue when parties are in crisis. How a crisis is handled directly relates to the probability of successful dispute resolution when Mediation is resumed.

Stress is the key element in crisis development. When an individual's normal level of functioning is interrupted by the occurrence of unusual stress due to single, multiple, or serial factors, the person will attempt to solve problems and handle tension in usual ways. If multiple attempts fail, a downward spiral of ineffective behavior, referred to as maladaptive behavior, occurs. As stress mounts to unusual proportions and the person's coping skills become increasingly ineffective, the potential for crisis increases.

Mediation cannot occur or continue in the face of crisis. The nature of crisis precludes any attempt by the parties to seek or explore alternatives or to problem solve. The party will look to the mediator to provide structure in a world that seems to be falling apart. It is at this point that the mediator's role is broadened to include crisis intervention. Crisis intervention is the act of interrupting the downward spiral as skillfully and quickly as possible and in so doing, returning the party to a pre-crisis level of functioning.

Because heightened stress is at the base of crisis, identifying some of the stressors attendant to a Mediation session would be helpful. These include

- seeing the other party.
- anticipating having to interact with the other party.
- being in the negotiation process.
- anticipating major changes in one or more areas of the party's personal and/or professional life.
- anticipating the loss of something significant in the party's life as a result of major changes. The losses may be categorized as follows:

  - Rituals:

    –Family reunions/celebrations
    –Sunday lunch/Saturday movie—weekly events that filled time and held sentimental value
    –Holiday traditions
    –Daily rituals—carpooling, athletic leagues

  - Titles/identity

    "The kids' teachers won't know who I am."
    "If I'm demoted, I'm nobody with nothing."
    "Being part of that family gave me status. It's about to end."

  - Relationships

    "Her parents were all the family I have here."
    "Our friends will side with him and I'll be alone."
    "This company has been my life."

  - Face

    "My parents will think I gave in to him."
    "The other guys on the shift are not going to believe I didn't just cave on this."
    "What if I lose today?"

  - Role/status

    "If she gets custody, I won't have a say in how the kids are raised. What if I can't make her see how important that is?"
    "I can't go from Chairman of the Board to
    _____."
    "I was their mentor. How will they see me when this is over? I've got to have that position!"

  - Form/structure

    "After this, everything might be changed. It's all mixed up."

"The predictability is gone. I can't function without structure."

- Hopes/expectations

  "I thought we would all grow old together."
  "I counted on the pension plan and the benefits to set me up for life."
  "We put everything into that building and it's gone. What if we don't get relief today?"

- Control

  "There is nothing I can do to reverse this."

- Security

  Fearing what might occur in the session. Concerns about an unknown outcome, personal performance, or mediator bias and skill sometimes create a high degree of free-floating anxiety in the parties as they approach a Mediation.

The above lists are not exhaustive. They serve only to alert the mediator to the need for awareness and preparation. To expect that the parties will have deposited all their feelings in a mythical "feelings bank" before entering the session is unrealistic.

A Proactive Approach to Minimizing the Disputants' Stress

1. Recognize the potential for heightened stress or emotional discomfort at any point prior to, during, or at the close of a Mediation session.
2. Arrange the office carefully. The structure of one's physical world contributes to increased stress. An alert mediator can eliminate or at least minimize these stressors. What follows is a list of potential environmental stressors within an office:

   - Glass offices that allow little privacy for their occupants
   - Open spaces where doors cannot be closed during times when privacy is needed
   - Windowless rooms
   - Insufficient insulation to limit sounds or conversations from another room
   - Artificial or natural lighting that produces glare
   - Inadequate heating or cooling
   - Ringing telephones or other distractions during the Mediation session
   - Overstuffed furniture that hampers free movement

- Clutter that produces a stifling effect
- Sharp, hot, dissonant wall colors that can be provocative or anxiety producing
- Intrusive background music
- Cigar, cigarette smoke
- Inadequate ventilation
- Inadequate work space
- A poorly planned seating arrangement that creates either too much or too little space between the parties and the mediator
- Wall treatments, window treatments, or furniture treatments that intrude
- Inadequate accommodations for the elderly or handicapped
- Inadequate security within or surrounding the office building

The list can be expanded. Mediators would do well to identify those areas in their own offices that can exacerbate the stress already being experienced by the parties when they enter the room. Such a search will additionally serve to minimize the mediator's stress.

3. Use the information gathered in the intake and through observation to optimize the room, seating, and safety arrangements and to determine the need to address special concerns in the introduction. Feelings might need to be addressed before moving into substantive issues.
4. Plan the opening statement.

- Take temporary control of the argument.
- Create a sense of security for the parties.
- Establish rapport, confidence, and trust with the parties.
- Set the stage, detail the process agenda, and explain Mediation. Eliminate confusion and misinformation.
- Address the stress often associated with conflict resolution.
- Legitimize the feelings expressed by the parties. Often the acknowledgment by the mediator of the presence of these feelings will serve to reduce the trauma being experienced. Be empathetic.
- Obtain clear commitments on ground rules.
- Answer questions.
- Be sensitive to concerns raised by the parties.

5. Caucus as appropriate (see chapter 6).
6. Empower the parties (see chapter 6).
7. Remain aware and sensitive throughout the session to what each party is communicating verbally and nonverbally.

Crisis Recognition

Careful planning in the various areas mentioned above and early response to the parties' rising stress are vital. Prevention and early management are the optimal situation. A more aggressive intervention is needed in situations where the stress continues to mount and the parties' coping skills are no longer effective.

Disputants may indicate crisis in different ways: crying out, exploding, verbalizing their suffering, or quietly withdrawing and shutting down. A person in crisis may evidence any of the following mental states that reflect possible crisis and result in maladaptive behavior. This may be characterized by their verbal as well as behavioral responses.

- Bewilderment: "I've never felt this way before."
- Danger: "I'm so nervous and frightened."
- Impasse: "I feel stuck. Nothing I do seems to help."
- Confusion: "I can't think clearly."
- Desperation: "I've just got to do something!"
- Apathy: "Nothing I do seems to help, so why bother anymore?"
- Helplessness: "I can't take care of myself."
- Urgency: "I need help now!"
- Discomfort: "I feel miserable. I'm restless and unsettled!"
- Anger: "How could this happen to me?"

Crisis that is exhibited verbally or in hostile, violent actions is easily recognized. The more subtle withdrawal or emotional shutting down is a challenge to the mediator. Withdrawal might be demonstrated by the following behaviors:

- A gradual turning of a shoulder or the total body away from the table

- A continuous blank stare
- A gradual nonparticipation in the process
- A flattened effect
- A sudden blanket agreement to everything being suggested

Procedure for Crisis Intervention

The following outlines procedures that the mediator should use in crisis intervention. Instructions are written for the mediator.

1. Immediacy. Intervention begins at the moment you recognize the person is in crisis. Attempt to relieve anxiety, prevent further disorientation, and ensure that sufferers do not harm themselves or cause harm to others.
   Key element: The mediation process ceases now, and the intervention begins. Remain in the room with the sufferer. Ask the other party to move to a caucus room. The co-mediator can manage this.
2. Control. Help reorder the chaos that exists in the person's world at the moment of crisis. Provide the needed structure and stability until the party is able to regain control.

   Mediator: Mrs. Jones, you're obviously experiencing a lot of discomfort.

   Mrs. Jones: I don't know what you or anybody else can do. I just don't know . . . I can't stand it anymore and I'm frightened.

   Mediator: Take a deep breath; hold it 5 seconds; let it out. Repeat. Repeat again. Is that helpful?

   Mrs. Jones: I feel a little calmer.

3. Assessment. Assess the situation. What is troubling the party now? Why did the person go into crisis at this particular time? What is necessary to implement the most effective help in the least time? Focus on the present crisis. What were the precipitating events?

   Mediator: Tell me what just happened, Bob.

   Bob: This is crazy! This whole thing is crazy!

Mediator: What is crazy, Bob?

Bob: I worked for this firm for 25 years. I am a good employee. I put in extra time when they asked. I showed up on time and did my job. I've been loyal and honest. The fact is I made my boss look good by what I produced.

Mediator: You have a fine record with the company.

Bob: Then why are they having me take a cut in salary and reducing my benefits?

Mediator: It sounds as though you're really hurting about all this.

Bob: You can't imagine the hurt I feel. This has shaken everything I believe in . . . I . . . I . . . I . . . I'm sorry I can't stop crying.

Mediator: Take your time.

Bob: I was taught that if you did a good job and were loyal . . . Everything I believed in is shattered. I can't think straight. I've got to do something to resolve the matter and end this pain, and I can't think straight today.

Mediator: What would be most helpful to you right now, Bob?

Bob: Just to say those feelings out loud. To tell somebody how bad it hurts.

Ask short direct questions.

Mediator: What are you feeling, Mr. Allison?

Mr. Allison: I feel lost and alone.

Mediator: What do you mean by "lost"?

Mr. Allison: It feels as though everything that's important is going to go away or be taken away after today.

Mediator: Is that the loneliness that you mentioned?

Mr. Allison: Yes. I've never been so scared in all my life. Even when I was a little kid.

Ask questions one at a time. Allow the party enough time to answer the questions.

Mediator: Did you ever feel this way before?

Jane: I can't think too clearly. That's what worries me.

Mediator: Take your time, Jane. We're in no rush.

Do not increase the party's confusion by bombarding him or her with many questions at once.

Learn to accept discomfort with silence. Recognize the usefulness of silence in the intervention process.

Make a statement describing how the behavior of the party is perceived.

Mediator: I can see you are very angry. (A sufferer will usually tell the intervener if the assessment is inaccurate.)

Ellen: No. It's not anger that I'm feeling. It's total frustration and disappointment with my whole world. I can't hold anything together any more . . . I want to reach a settlement today, but I can't seem to concentrate.

Interrupt the party judiciously. Use periodic interruptions to clarify, to check the accuracy and your understanding of the person's statements, and simply to remind the party of your interest in the problem. Interrupt no more often than absolutely necessary.

Clarify the crisis.

Mediator: You're afraid that if the situation isn't resolved here today, the home owners will call a press conference and publicize the whole matter.

Contractor: Right. If this gets in the news, not only will I be ruined but my family's reputation will be ruined. This is an old family business. My dad trusted me. I can't handle the embarrassment.

Allow the crisis to be the party's crisis. Avoid judgments, preaching, and put-downs. The way the party currently perceives the world is the party's reality.

Assess both the actual and the symbolic meaning of the crisis event. Remember that perception triggers crisis much more often than do facts.

Use the party's body language and nonverbal language as a source of information. If you observe that the words and behavior do not match, question the discrepancy.

Listen for what is not being said.

Recognize that your personal attributes contribute to overall effectiveness.

- Remain calm and reassuring.
- Remain empathetic and attentive.
- Remain supportive.
- Be willing to reach out to the party, both emotionally and physically, as needed.
- Maintain a caring attitude that conveys a willingness to listen.

Allow the party to speak freely and ventilate feelings

4. Decide how to handle the situation after assessing it. Heightened stress closes down options and generally produces tunnel vision in a sufferer. When effective intervention occurs, the party becomes more receptive to exploring options, thinking creatively, and solving problems.

5. Refer and follow up as needed.

Assess the possibility of continuing with the Mediation. Determine whether the parties can resume the current session, whether it would be more prudent to reschedule for another time, or whether the Mediation should be terminated.

CHAPTER TWELVE

Practical Tips for the Mediator

The following are additional practical tips to enhance the mediation process.

1. Take personal safety and the possibility of weapons seriously. If the concern for safety factors in a mediation session becomes as much a second nature to the mediator as any other negotiation skills, the likelihood of positive outcomes is increased.

   Mediator Safety Guidelines
   Do not take the possibility of weapons lightly.
   Plan in advance for your personal safety and for the safety of clients.
   Check the arrangement of your office or mediation room. Be sure that all safety factors are accounted for before beginning a mediation session.
   When you greet clients, notice anything strange or unusual about their words, actions, or dress. Allow your senses to give you clues, and take all clues seriously until you have been able to rule them out. Probably your best tool to detect potential violence or the presence of weapons is your own internal communications and sensing capabilities.
   Read the participants' body language as they enter.
   Avoid having your back to clients, especially those who are suspected of having weapons or those who may erupt violently.
   Enter the Mediation room behind clients.
   Have clients sit with their back in the direction of the door through which they will go when they leave.
   If possible, arrange several seating areas within the room that can be used as needed with different threat levels.

Remove any potential weapons from the mediation room and from the table or desk used for Mediation prior to bringing in the parties. Recognize that paperweights, scissors, letter openers, mugs, and desk clocks are potential weapons.

Plan personal attire. Recognize that loop earrings, chain necklaces, ties, and scarves are easy targets for hostile clients.

Seat clients so that the potentially violent person does not have to pass the other party when entering the room.

Use the table or desk as a barrier between mediator and client or between client and client when needed.

Sit in a manner that conveys openness and interest without compromising the ability to react quickly to outbreaks of violence. Expecting the unexpected means being ready and able to respond as needed.

Attempt to keep the parties seated during the mediation session. It is much harder to take an aggressive stand from a sitting position than from a standing position.

Discuss suspected weapons with each party separately. Use a direct, precise, and noneuphemistic approach.

Do not remain in the office after hours with a potentially violent client unless proper security is available.

Arrange a "buddy system" so that someone is available should you need help.

Discuss and establish safety procedures with all staff members—receptionists, secretaries, mediators, agency director.

Know your emergency numbers and use them if you need help.

If at all possible, have a co-mediator.

Make contingency plans for all Mediations. Consider what you would do under a variety of conditions. Learn to play the "what if" game. Preplanning will affect automatic behavior under stress.

Take seriously your safety and the safety of your clients. Refuse to mediate if your concerns about safety are not satisfied.

At the completion of a hostile or emotional Mediation, arrange for each party to leave separately.

Debrief with the co-mediator. Be in touch with your feelings about what took place.

2. Take notes judiciously during the mediation session:
   a. Explain to the parties that all notes taken in the session will be destroyed at the end of the Mediation.
   b. Encourage disputants to take notes. Provide them with paper and pen to note their observations, corrections, ideas, and questions as a way to avoid interrupting the speaker. Assure them that they will be given their turn to speak.
   c. Write only as much as absolutely necessary. The mediator is expected to actively listen and respond. It is a difficult task made more difficult if the mediator is preoccupied writing. Limit notes to names/seating plan of those present, key statements of wants/needs, compelling words or phrases, statements requiring clarification, offers, and proposals.
   d. Put your pen down on the table between notes. The parties sometimes view the pen as a barrier between themselves and the mediator.

   Mr. Jones Mediator, please stop your writing and listen to me. You're just like Jane. She's always doing something else when I'm trying to talk to her.

   e. At the conclusion of the Mediation, collect and destroy notes if there has been a previous agreement to do so or if that is a procedural requirement.
3. Don't believe everything that is said by the parties. Disputants will often posture, overestimate, or underestimate. If something does not sound right, ask for clarification. Question inconsistencies for corrections without being combative, punitive, or positional.
4. As long as the parties remain at the table, resolution is possible. Their presence means they want or need something from the process. Make that need an advantage. The parties' behavior tells you more about what they want than do their words.
5. Change the disputant's perspective. Help the party see dif-

ferent points of view rather than trying to change his or her current point of view.

6. Be aware that some offers will inhibit or even stop movement toward settlement. Disputants will reject offers that do not reflect a concern for their interests or needs.

7. Allow the parties to do some venting. Observe the process for clues. Listen for key words, expressions, feelings. Listen for the content of the message and the meaning behind the content. Allowing the venting and observing the parties during the venting may provide the mediator with valuable information concerning actual needs and the style of interaction between the parties.

8. Don't spend time on nonproductive issues. Focus on current issues over which the disputants have control.

9. Attend to the importance placed on an issue by the disputant. Listen for content, feelings, and point of view. Note extra emphasis the party places on certain words. Watch for nonverbal clues as the party speaks.

10. Assess movement gained from bad offers as well as good offers.

11. Listen for the soft spots in a disputant's case and arguments. Soft spots are those parts of the arguments that don't make sense, are not related to objective criteria, or are related to some issues in the person's life other than what is on the table.

REFERENCES

Black, H. (1983). *Black's law dictionary* (5th ed.). St. Paul, MN: West.

Evarts, W. R. (1980). (transparency) Dallas: Dispute Mediation Service.

Evarts, W. R., Greenstone, J. L., Kirkpatrick, G. J., & Leviton, S. C. (1983). *Winning through accommodation: The mediator's handbook.* Dubuque, IA: Kendall/Hunt.

Greenstone, J. L., Leviton, S. C., & Fowler, C. M. (1991). *A mediation primer.* Unpublished manuscript.

Greenstone, J. L., & Leviton, S. C. (1993). *Elements of crisis intervention: Crises and how to respond to them.* Pacific Grove, CA: Brooks/Cole.

Moore, C. W. (1986). *The mediation process: Practical strategies for resolving conflict.* San Francisco: Jossey-Bass.

Texas Civil Practice and Remedies Code (1988 ed.). St.Paul, MN: West.

Zaidel, S.R. (1990). *Divorce with respect: A guide to divorce and divorce mediation in Israel.* Tel-Aviv, Israel: OR'AM Publishing House, Ltd.

BIBLIOGRAPHY

Coogler, O. J. (1978). *Structured mediation in divorce settlements.* Lexington, MA: Lexington Books.

Erickson, S. K., & Erickson, M. S. (1988). *Family mediation casebook: Theory and process.* New York: Brunner/Mazel.

Evarts, W. R., Greenstone, J. L., Kirkpatrick, G., & Leviton, S. C. (1983). *Winning through accommodation: The mediator's handbook.* Dubuque: Kendall/Hunt.

Fisher, R., & Ertel, D. (1995). *Getting ready to negotiate.* New York: Penguin.

Fisher, R., & Ury, W. (1981). *Getting to yes: Negotiating agreement without giving in.* Boston: Houghton Mifflin.

Folberg, J., & Taylor, A. (1984). *A comprehensive guide to resolving conflict without litigation.* San Francisco: Jossey-Bass.

Fowler, W. R., & Greenstone, J. L. (1983). Hostage negotiations. In R. Corsini (Ed.), *Encyclopedia of psychology* (p. 142). New York: Wiley.

Fowler, W. R., & Greenstone, J. L. (1987). Hostage negotiations for police. In R. Corsini (Ed.), *Concise encyclopedia of psychology* (pp. 530–531). New York: Wiley Interscience.

Friedman, G. J. (1993). *A guide to divorce mediation: How to reach a fair, legal settlement at a fraction of the cost.* Workman Publishers.

Goldberg, S. B., Green, E. D., & Sander, F. E. A. (1985). *Dispute resolution.* Boston: Little, Brown.

Greenstone, J. L. (1978). An interdisciplinary approach to marital disputes arbitration: The Dallas plan. *Conciliation Courts Review, 16,* 7–15.

Greenstone, J. L. (1992). The key to success for hostage negotiations teams: Training, training, and more training. *The Police Forum, 1,* 3–4.

Greenstone, J. L. (1992, April). *Mediation advocacy: A new concept in the area of family dispute resolution.* Paper presented at the meeting of the 6th International Congress on Family Therapy: Divorce and Remarriage Interdisciplinary Issues and Approaches, Jerusalem, Israel.

Greenstone, J. L., & Leviton, S. C. (1981). Crisis management and intervener survival. In R. Corsini (Ed.), *Innovative psychotherapies* (pp. 216–228). New York: Wiley Interscience.

Greenstone, J. L., & Leviton, S. C. (1982). *Crisis intervention: Handbook for interveners.* Dubuque: Kendall/Hunt.

Greenstone, J. L., & Leviton, S. C. (1983). Crisis intervention. In R. Corsini (Ed.), *Encyclopedia of psychology* (pp. 312–314). New York: Wiley.

Greenstone, J. L., & Leviton, S. C. (1983, March). *Mediation: An alternative to litigation.* Paper presented at the meeting of the Academy of Criminal Justice Sciences, San Antonio, TX.

Greenstone, J. L., & Leviton, S. C. (1983, July). *Mediation: Family dispute resolution.* Paper presented at the meeting of the 4th International Congress of Family Therapy, Tel Aviv, Israel.

Greenstone, J. L., & Leviton, S. C. (1984, March). *Divorce mediation: The way of the 80's.* Paper presented at the meeting of the Oklahoma Association of Marriage and Family Therapists, Tulsa, OK.

Greenstone, J. L., & Leviton, S. C. (1984, September). *Management mediation: The police officer's alternative to litigation.* Paper presented at the meeting of the First National Symposium on Police Psychological Services, Quantico, VA.

Greenstone, J. L., & Leviton, S. C. (1984, December). *Crisis management for the mediator.* Paper presented at the meeting of the Second Annual Conference on Problem Solving Through Mediation, Albany, NY.

Greenstone, J. L., & Leviton, S. C. (1986). Mediation: The police officer's alternative to litigation. In *Psychological services for law enforcement.* Washington, D. C.: U.S. Department of Justice, Federal Bureau of Investigation, U.S. Government Printing Office.

Greenstone, J. L., & Leviton, S. C. (1986, June). *Alternatives in dispute resolution: Family and marital mediation.* Paper presented at the meeting of the 5th International Congress of Family Therapy, Jerusalem, Israel.

Greenstone, J. L., & Leviton, S. C. (1986, July). *The dispute mediator as crisis manager: Crisis intervention skills for the mediator in high stress, high risk situations.* Paper presented at the meeting of the Academy of Family Mediators, Minneapolis, MN.

Greenstone, J. L., & Leviton, S. C. (1987). Crisis management for mediators in high stress, high risk, potentially violent siutations. *Mediation Quarterly, 3,* 20–30.

Greenstone, J. L., & Leviton, S. C. (1987, July). *Crisis intervention for mediators in high risk, high stress, potentially violent situations.* Paper presented at the meeting of the Academy of Family Mediators, New York, NY.

Greenstone, J. L., & Leviton, S. C. (1993). *Elements of crisis intervention.* Pacific Grove, CA: Brooks/Cole.

Haynes, J. M. (1994). *The fundamentals of family mediation.* New York: SUNY Press.

Kolb, D. M. (1983). *The mediators.* Cambridge, MA: MIT Press.

Kressel, K., Pruitt, D. G., & Associates. (1989). *Mediation research: The process and effectiveness of third party intervention.* San Francisco: Jossey-Bass.

Lemmon, J. (1985). *Family mediation practice.* New York: Free Press.

Leviton, S. C. (1983). Conflict mediation. In R. Corsini (Ed.), *Encyclopedia of psychology.* New York: Wiley.

Leviton, S. C., & Greenstone, J. L. (1983, March). *Divorce mediation and the attorney.* Paper presented at the meeting of the Family Law Section of the Dallas Bar Association, Dallas, TX.

Marlow, L., & Sauber, S. R. (1990). *The handbook of divorce mediation.* New York: Plenum.

Pederson, P. A. (1988). *A handbook for developing multiculture awareness.* New York: American Association for Counseling and Development.

Raiffa, H. (1982). *The art and science of negotiation.* Cambridge, MA: Harvard University Press.

Rogers, N. H., & Salem, R. A. (1987). *Student's guide to mediation and the law.* New York: Matthew Bender.

Singer, L. R. (1990). *Settling disputes: Conflict resolution in business, families, and the legal system.* Boulder, CO: Westview.

Suskind, L., & Crulkshank, J. (1987). *Breaking impasses.* New York: Basic Books.

Ury, W. (1991). *Getting past no.* New York: Bantam Books.

Ury, W. L., Brett, J. M., & Goldberg, S. B. (1983). *Getting disputes resolved.* San Francisco: Jossey-Bass.

TO THE OWNER OF THIS BOOK:

We hope that you have found *Elements of Mediation* useful. So that this book can be improved in a future edition, would you take the time to complete this sheet and return it? Thank you.

School and address: _____

Department: _____

Instructor's name: _____

1. What I like most about this book is: _____

_____

_____

2. What I like least about this book is: _____

_____

_____

3. My general reaction to this book is: _____

_____

4. The name of the course in which I used this book is: _____

_____

5. Were all of the chapters of the book assigned for you to read?

    Yes   No

If not, which ones weren't? _____ `

6. On a separate sheet of paper, please write specific suggestions for improving this book and anything else you'd care to share about your experience in using the book.

Optional:

Your name: _____ Date: _____

May Brooks/Cole quote you either in promotion for *Elements of Mediation* or in future publishing ventures?

Yes: _____ No: _____

Sincerely,

Sharon C. Leviton
James L. Greenstone

FOLD HERE

‖‖‖‖

BUSINESS REPLY MAIL
FIRST CLASS   PERMIT NO. 358   PACIFIC GROVE, CA

POSTAGE WILL BE PAID BY ADDRESSEE

ATT: _Sharon C. Leviton & James L. Greenstone_

Brooks/Cole Publishing Company
511 Forest Lodge Road
Pacific Grove, California 93950-9968

‖ı‖ıııı‖ı‖ı!ıııı‖ı‖‖ıııı‖ı‖ıı‖ı‖ııı‖‖ı‖ı‖ıı‖ı‖

FOLD HERE

NOTES